# ACRI '96

Springer-Verlag London Ltd.

S. Bandini and G. Mauri (Eds)

# ACRI '96

**Proceedings of the Second Conference on Cellular
Automata for Research and Industry,
Milan, Italy, 16-18 October 1996**

With 79 figures

 Springer

S. Bandini and G. Mauri

Department of Computer Science, University of Milan, Milan 20135, Italy

Cover based on a design by L. Moroni and S. Cesaroni

ISBN 978-3-540-76091-7        ISBN 978-1-4471-0941-9 (eBook)
DOI 10.1007/978-1-4471-0941-9

British Library Cataloguing in Publication Data
Conference on Cellular Automata for Research and Industry (2nd : 1996 : Milan, Italy)
   ACRI '96 : proceedings of the second Conference on Cellular Automata for Research and
   Industry, Milan, Italy, 16-18 October 1996
   1.Cellular automata - Congresses
   I.Title   II.Bandini, Stefania   III.Mauri, Giancarlo   IV.Conference on Cellular Automata for
   Research and Industry (2nd : 1996 : Milan, Italy)
   004.3'5

Library of Congress Cataloging-in-Publication Data
A catalog record for this book is available from the Library of Congress

Typesetting: Camera ready by editors

34/3830-543210 Printed on acid-free paper

# Foreword

ACRI'96 is the second conference on Cellular Automata for Research and Industry; the first one was held in Rende (Cosenza), on September 29-30, 1994.

This second edition confirms the growing interest in Cellular Automata currently present both in the scientific community and within the industrial applications world.

Cellular Automata-based computational models, besides capturing the attention of scientists working in different fields, open new perspectives of intersection between different and historically distant areas of scientific knowledge, from Physics to Biology, to Computer Science.

ACRI'96 aims at providing a forum both for researchers working in the Cellular Automata field, and for those who foresee the possibility to verify on concrete domains of application the impact of their solutions, as well as for those who are looking for a possibility of reflection upon the specific concept of parallel and distributed computation provided by Cellular Automata.

This book contains the works presented at the conference. The invited papers cover different aspects of Cellular Automata. T.Worsch gives a classification of Cellular Automata mapping on the existent computational frameworks for the simulation of their behavior. One of the most mature areas where Cellular Automata showed their value is Physics: B.Chopard illustrates recent results on wave modeling, and some possible applications. According to the general purpose of ACRI'96 of transfering research results to the industrial world, F.Forlani presents the point of view of an industry top manager on the role of the approach based on Cellular Automata in defining core competency for industrial innovation, especially regarding applications involving knowledge about complex systems.

The contributed papers cover a wide range of interdisciplinary topics spanning from theoretical aspects of Cellular Automata to applications to real problems. In particular, topics treated include:

- Theoretical aspects of CA: Chaos related issues [Cattaneo et al., Finelli et al.], Periodic collective behavior [Jimenez-Morales et al.], Multilayered automata networks [Bandini et al.];
- Applications to biological problems: Reconstruction of cellular dynamics [Bignone], Calcium-ion distribution in the living cell [Fesce et al.], Prey-predator systems [Bandini et al.];
- Applications to problems in Physics: Water flow through a porous soil [Di Gregorio et al.], Diffusion [Castagnolo et al.];
- Applications to Urbanistics [Rabino et al.];
- Other applications: Solution of routing problems [Hochberger et al.], Image processing techniques [Adorni et al.], Associative memories [Pessa et al.];
- Computational environments and programming languages for Cellular Automata: [Sipper et al., Spezzano et al., Hartmann et al.].

We would like to thank all the Scientific Committee members for their involvement as referee and all the participants.

Special thanks go to all the people who have helped in the organization of the conference, especially to S.Betelli, T.Bezzi, R.Casati, E.Formenti, and M.Magagnini.
S.Cesaroni and L.Moroni deserve our gratitude for designing and taking care of all the graphics of the conference.
Finally, we gratefully acknowledge support from AEI, AIS, CNR, Eniricerche, Illycaffè, Consorzio Milano Ricerche, Philips Automation, Pirelli Tires, and the University of Milan.

Stefania Bandini
Giancarlo Mauri

CONFERENCE CHAIRMAN
Giancarlo Mauri (University of Milan)

INVITED LECTURES
Thomas Worsch (University of Karlsruhe)
Bastien Chopard (University of Geneva)
Franco Forlani (Italian Association for Industrial Research)

SCIENTIFIC COMMITTEE
Stefania Bandini (University of Milan)
Giampietro Boniardi (Philips Automation S.p.A, Monza)
Gianpiero Cattaneo (University of Milan)
Bastien Chopard (University of Geneva)
Vincenzo D'Andrea (University of Parma)
Salvatore di Gregorio (University of Calabria)
Riccardo Fesce (CNR - DIBIT - H. S. Raffaele, Milan)
Mario Furnari (Institute of Cybernetics - Arco Felice)
Giancarlo Mauri (University of Milan)
Jacques Mazoyer (LIP-ENS - Lyon)
Patrizia Mentrasti (University "La Sapienza" of Rome)
Mario Molinari (Eniricerche S.p.A. - S. Donato Milanese)
Giovanni Rabino (DISET - Politecnique of Milan)
Paola Rizzi (STRATEMA-DAEST - Venice)
Roberto Serra (Montecatini Environmental Research Center - Ravenna)
Giandomenico Spezzano (University of Calabria)
Furio Suggi Liverani (Illycaffè S.p.A. - Trieste)
Francesco Tisato (University of Milan)
Marco Tomassini (Ecole Politechnique Federale - Losanne)
Giuseppe Trautteur (University of Neaples)
Thomas Worsch (University of Karlsruhe)

ORGANIZING COMMITTEE
Stefania Bandini (University of Milan)
Marco Bistolfi (Eniricerche S.p.A. - S. Donato Milanese)
Roberto Casati (University of Milan)
Enrico Formenti (University of Milan)
Lorenzo Papini (Politecnique of Milan)

# Contents

## Invited Lectures

## Contributed Papers

# Invited
# Lectures

# Programming Environments for Cellular Automata

Thomas Worsch

Department of Informatics, University of Karlsruhe

76128 Karlsruhe

worsch@ira.uka.de

### Abstract

Some modifications and generalizations of cellular automata are discussed which are sometimes useful in the modeling of real phenomena and which therefore have found their ways into some programming environment for cellular automata. In the second part several aspects are discussed with respect to which these programming environments can be compared.

## 1 Introduction

### 1.1 Why CA Simulations?

It is reasonable to use computers whenever computations have to be done[1] and the number and/or complexity of the operations involved exceeds what can reasonably be done by a human being. One possible cause of this is a large amount of data items to be processed. Sometimes it is also the large amount and the required special form of outputs which make computers indispensable, e.g. when showing the animation of an algorithm to a class of students.

Since these circumstances arise quite naturally in connection with cellular automata (CA), quite a lot of software packages (SP or simply "system" hereafter) for CA have been designed and made available.

Such SP typically require the local transition rule(s) to be specified. The SP are therefore obviously suitable for forward problems ("from local to global"). But sometimes inverse problems ("from global to local") are (or even can only be) tackled by searching for appropriate local transition rules which makes the SP useful in this case, too.

Instead of writing new programs each time which are similar to the earlier ones, it seems useful to provide environments which release the user from doing the same things again and again. Furthermore it is possible for a SP to provide better solutions for certain subtasks than the user would have found herself.

---

[1] It is not reasonable to use them, if their usage can be avoided by a theoretical argument; and they cannot be used as an equal replacement for the latter.

Most SP offer *special programming languages* for the specification of CA, most notably their local transition functions. These languages are the topic of interest in this paper.

## 1.2 About This Paper

In section 2 the basic definitions for CA and some of their modifications and generalizations are reviewed.

In section 3 some programming language approaches to the specification of CA are discussed.

Due to a lack of space neither the possibilities offered by the different SP for describing the interaction between the simulated CA and its environment (running simulations, IO) will be discussed nor the details of several available SP. These can be found in the long version of the full paper, to appear as a technical report [23]. It will be available from the WWW page http://liinwww.ira.uka.de/~worsch/papers/index.html#prog-envs.

Further information about software packages (including URLs to some software repositories) will be accessible on the WWW page http://liinwww.ira.uka.de/~worsch/ca/prog-envs.html.

Since the aim of this paper is not proving anything only a few formal definitions will be given. Readers interested in theoretical aspects and results can always refer to the indicated literature.

# 2 Cellular Automata

## 2.1 Standard CA

A *cellular automaton* is usually specified by a *lattice L* of cells, a finite set of *states Q* for the cells, a finite *neighborhood N*, and a *local transition function* $\delta : Q^N \rightarrow Q$. [2]

In the case of generalizations of the standard model to be discussed later $\delta$ sometimes isn't a function any longer and we will therefore sometimes more generally speak of a *local transition rule*.

The regularity of $L$ can be formalized using Cayley graphs (i.e. $L$ is a group). In the available SP considered usually only Euclidean grids $L = \mathbf{Z}^d$ are allowed. A *global configuration* is a mapping $c : L \rightarrow Q$ assigning a state to each cell.

In standard CA the neighborhood $N = \{n_1, \ldots, n_k\}$ is the same for all cells all of the time. The neighbors of a cell $l \in L$ are $l + n_1, \ldots, l + n_k$. A mapping $d : N \rightarrow Q$ is called a *local configuration*. The local transition rule maps local configurations to states.

In computer simulations the situation becomes a little bit more complicated, because infinite lattices $L$ may pose storage and computation time problems. Restricting the CA to some finite lattice may violate its regularity. Therefore we introduce a so-called *localization function* $\lambda : L \times Q^L \rightarrow Q^N$. For a cell $l$ and a global configuration $c$ it describes what should be considered the local configuration $\lambda(l, c)$ at $l$ in $c$. [3]

---

[2] We are using the notation $B^A$ for the set of all functions from $A$ to $B$.

[3] In general $\lambda$ must obey some reasonable restrictions of course.

In the case of finite lattices like $L = \{0, \ldots, m\}^d$ it is possible to model tori as well as grids with default state value "outside" of the grid using some appropriate localization functions. For example for a two-dimensional torus one could define $\lambda((0,0), c)((-1,0)) = c(m, 0)$. To put it in a different way, the localization function can be used to model what is sometimes called the *boundary conditions*.

The *global transition function* $\Delta : Q^L \to Q^L$ (or the global transition relation $\vdash_{\delta, det}$) induced by $\delta$ and $\lambda$ is defined as

$$\Delta(c) = c' \iff c \vdash_{\delta, det} c' \iff \forall l \in L : c'(l) = \delta(\lambda(l, c)) \tag{1}$$

Choosing $L = \mathbf{Z}^d$ and $\lambda(l, c)(n) = c(l + n)$ results in the classical type of CA as described by von Neumann.

In the case of lattices $L = \mathbf{Z}^d$ the neighborhood $N_r = \{(x_1, \ldots, x_d) \mid \forall i : 0 \le x_i \le r\}$ is called the Moore neighborhood with radius $r$.

Let us call a CA satisfying the above requirements a *standard CA*.

## 2.2 Modifications and Generalizations of the Standard CA Model

There are several modifications and extensions of standard CA which have been considered in the literature, which are also useful for some modeling purposes, and which are supported by some SP.

We use the following nonstandard terminology. By *modifications* of CA we mean computational models the definitions of which differ from those for CA but which can simulate CA and which can be simulated by CA with an only linear overhead in time and number of cells. On the other hand when we speak of a *generalization* of CA we mean a model which at least cannot be shown to be simulated by CA in linear time. So, in some sense a generalization is more powerful than standard CA, while a modification (in this paper at least) "only" presents a (more or less) different view on the same subject. [4]

The following list should neither be considered complete nor should one expect the different aspects discussed to be independent. It sometimes depends on the point of view whether one categorizes a certain aspect in way or the other.

### 2.2.1 Partitioned CA

In a standard CA a cell may use the whole state, in other words *all* bits, of each neighboring cell in order to determine its own new state. In a *partitioned CA*[5] (in the sense of [14]) this is no longer the case. Instead the state set is of the form $Q = Q_{n_1} \times \cdots \times Q_{n_k}$ where $N = \{n_1, \ldots, n_k\}$. The local transition function is now written as $\delta : Q_{n_1} \times \cdots \times Q_{n_k} \to Q_{n_1} \times \cdots \times Q_{n_k}$. Correspondingly the range of $\lambda$ is now $Q_{n_1} \times \cdots \times Q_{n_k}$, too, and it has the meaning that from the state of a neighbor $n_i$ a cell only "reads" the state component $Q_{n_i}$.

Besides interesting theoretical properties, partitioned CA also have an advantage from a practical point of view: The the domain of $\delta$ in some sense is smaller: instead

---

[4] The notion of *simulation* is used here in an informal way only.

[5] Readers should be aware of the fact that the similar notion of *partitioning CA* is also in use. This will be covered in the subsubsection on "Block rules".

of size $|Q|^{|N|}$ it only has size $|Q|$. This may make possible the implementation of a $\delta$ which otherwise would be infeasible. A typical example is the CAM-8 machine. There the states of the cells may comprise at most 16 bits. Assume that this is the case and that the two-dimensional radius 1 Moore neighborhood is used. In a standard CA this would mean that $\delta$ has $9 \cdot 16 = 144$ bits as arguments. In the CAM-8 the local transition function is stored as a full lookup table, which in this case would consist of $2^{144} \cdot 16$ bits. The CAM-8 only allows – but often this is enough – partitioned CA; each bit of a state may be used only by *one* (other) cell. Therefore $\delta$ only uses 16 input bits to compute a new state and the lookup table only consists of $2^{16} \cdot 16$ bits.

### 2.2.2  Block Rules

CA using *block rules* are also called *partitioning CA*. (This should not be confused with partitioned CA, although there are some connections.) Their specific features are a modified use of the neighborhood and a slightly different type of the local transition function corresponding to it.

A lot of algorithms for CA can be described quite easily using the notion of *signals*. They can be specified in standard CA, as can already be seen for example from the construction by von Neumann [20].

But this doesn't mean that it's the best framework, too. In this and the following subsubsection two modifications of standard CA will be discussed which both allow an easier definition of signals (or other "moving entities").

We'll describe only the most prominent example type of block rules, i.e. CA using the so-called Margolus neighborhood [19]. For the sake of simplicity consider the lattice $L = \mathbf{Z}^2$ and two tilings $T_1$ and $T_2$ of $L$. Both consist of copies of a $2 \times 2$ "neighborhood blocks" $N$. In $T_1$ the upper left corners of the tiles are those cells for which both coordinates are even, and in $T_2$ the upper left corners of the tiles are those cells for which both coordinates are odd.

The block rule, i.e. the local transition function for a partitioning CA with Margolus neighborhood has the form $\delta : Q^N \rightarrow Q^N$. Starting from an initial configuration in the first, third, fifth, ... transition step $\delta$ it is applied to all blocks of $T_1$ simultaneously, while in the second, forth, sixth, ... step it is applied to all blocks of $T_2$.

In the one-dimensional analogue the blocks are copies of $N = \{0, 1\}$. In this case it is easy to define a block rule for a horizontally moving 1 in a "sea" of 0's. One simply defines $\delta(\boxed{10}) = \boxed{01}$. This is easier to understand than a signal in standard CA.

For a more general treatment of partitioning CA see [6].

### 2.2.3  Automata Moving Around on a CA Lattice

One can go one step further in the modification of CA resulting in models which can be "programmed" even more conveniently. Although sometimes attributed to other authors the idea for these kinds of models at least goes back to Hemmerling [9].

*Parallel Turing machines* (PTM), introduced by Wiedemann [22], are a generalization of Turing machines where on the tape(s) instead of one finite automaton several may be moving around which in addition to reading and writing the tape can also generate additional finite automata and delete themselves. One-tape PTM can be regarded as a modification of one-dimensional CA.

Substituting the "dead" tape by a CA again only is a formal modification; the result is still a model which can simulate and can be simulated by CA trivially.

In this framework for example a signal moving to the right could be easily specified as an appropriate finite automaton.

### 2.2.4  Asynchronous CA

The local transition rule of an *asynchronous CA* is specified as a function $\delta : Q^N \to Q$, but its meaning is not the same as for standard CA. Actually one can define:

$$c \vdash_{\delta, asNnc} c' \iff \forall l \in L : c'(l) \in \{\delta(\lambda(l,c)), c(l)\} \tag{2}$$

In an asynchronous CA each cell can in each step choose nondeterministically between changing its state according to $\delta$ or keeping its current state.

For more information on asynchronous and more generally nondeterministic CA, readers are referred to [8] and the references mentioned there.

### 2.2.5  Probabilistic CA

*Probabilistic CA* are different from nondeterministic CA. Although a cell may enter different new states $q \in Q$ if it observes a certain local configuration $d$ in its neighborhood (as in the nondeterministic case) this time for each $q$ a probability $\delta(d, q)$ is specified with which it is entered. Hence a local transition rule is of the form $\delta : Q^N \times Q \to [0; 1]$ satisfying for all $d \in Q^N$ the restriction that $\sum_{q \in Q} \delta(d, q) = 1$.

Although in probabilistic CA (as well as nondeterministic ones) it is meaningless to speak about *the* result of *the* computation for a certain initial configuration of a CA, interestingly enough there are CA models for phenomena observed in nature which are probabilistic. For example in lattice gases certain local configurations may lead to two different new states of the center cell with equal probability.

### 2.2.6  Inhomogeneous CA

Since CA are homogeneous in several ways (lattice, neighborhood, local transition rule) there are also several possibilities to break homogeneity. We subsume all of these approaches under the notion of *inhomogeneous CA*. Up to now there are only few publications (e.g. [13, 15]) considering such aspects.

It is useful to distinguish between temporal and spatial inhomogeneity (which do not exclude each other).

An example for temporal inhomogeneity would be the following (taken from one of the typical applications for CA): Assume that one wants to processes images the pixels of which are stored in the cells of a CA. Assume further that the processing consists in the application of several filters or other operators to an image, and each of the operators is implemented as the local transition rule of a CA. Then it obviously would be helpful if one could specify a generalized CA where first for some number of steps the local rule implementing the first operator could be applied, afterwards to the result of this, i.e. the contents of the cells of the CA, for some more steps the local rule implementing the second operator, and so on. In other words the transition function for the cells would (be allowed to) change during time.

Probably the simplest example of a CA with spatial inhomogeneity is a standard CA on say a two-dimensional Euclidean lattice which one wants to restrict to a finite rectangle. If one uses for example Moore neighborhood then there is a problem with the cells which are placed at the borders of the lattice. Instead of using an appropriate localization function $\lambda$ as introduced in subsection 2.1 one can allow that the size and the shape of the neighborhood of a cell may vary throughout the lattice, and similarly the local transition rule used by the cells.

If one wants to investigate inhomogeneous CA from a computation theoretic point of view, it is of course necessary to impose certain restrictions upon the model. Its inhomogeneity must be computable in some sense. But we won't go into further details here (see the papers cited above).

Finally there is yet another aspect of generalization which could also be considered a manifestation of inhomogeneity but because of its importance for SP is discussed separately.

### 2.2.7  Hierarchical Aspects of Generalized CA

Form the theoretical point of view there are only very few papers dealing with hierarchy [1, 5]. What we would like to subsume under the notion of hierarchy are approaches where the complete cells of a CA are composed from simpler parts somewhat analogous to semidirect products in algebra. In other words there are (noncyclic) dependencies different "parts" of a state when computing the new state of a cell. From a global point a view this corresponds to a set of CA which are partially ordered.

## 3   CA Programming Language Concepts

For the simulation of CA on a computer there are at least two basic possibilities. One is to offer a library of routines to be used with a general purpose language, e.g. [12]. One possible disadvantage of this approach is that the compiler does not know anything about the special application and hence cannot do any specific optimizations. This can be overcome by the second approach which is to offer a special "CA programming language". [6]

Not all systems offer the same degree of flexibility (concerning e.g. the dimensionality or the size of the lattice) and they also differ with respect to the extensions of standard CA which can be used within them. Therefore some of the aspects mentioned in the previous section are reconsidered with respect to some CA software packages and CA programming languages which are available. These are Camel/Carpet [18], CAM Simulator, CANL [4], capow, CaSim, CAT/CARP, CDL [11], CDM/SLANG [17], Cellas/Fundef, cellsim, Cellular/Cellang [7], Ceprol [16], DDLab, Hical [3], LCAU, scamper/cal, Scarlet/SDL, and many others. [7] (For an earlier short survey see [10].)

---

[6]More precisely one should probably speak of a notational system for the specification of CA.

[7]If two names A/B are given, A denotes a whole CA system and B the programming language of A.

## 3.1 Lattice

Structure.    DDLab allows arbitrary connections between the cells (*random networks*). The authors of most other programming environments say, that the lattice is a, most of time *two-dimensional, Euclidean* one; Cellang and SDL do not impose any restrictions on the number of dimensions. The former usually means that (relative) coordinates consisting of two components can be used to specify a neighbor. But one should observe that, as long as it is possible to use arbitrary neighborhoods, is it also possible to "embed" for example a finite three-dimensional grid into a finite two-dimensional one. For example in order to simulate a $k \times k \times k$ cube with six nearest neighbors for each cell, one uses a $k \times k^2$ grid where the six neighbors of a cell are at the positions $(\pm 1, 0)$, $(0, \pm 1)$ and $(0, \pm k)$. Admittedly this is a bit awkward.

Size.    In most programming environments the lattice size is only bounded by the amount of available free memory. At least Cellular does some optimizations in the case where the side lengths of the grid are a power of two. CANL requires the grid to be a square.

Boundary Conditions.    Most packages allow the user to choose at least between *fixed* boundary conditions and *toroidal* boundary conditions. Some packages allow this to be specified separately for the different dimensions. CANL also allows so-called *adiabatic* boundary conditions, where a non-existent neighboring cell is assumed to have the same state as the center cell. CaSim allows so-called *reflecting* boundary conditions, where the finite lattice is virtually extended by repeatedly mirroring it at its borders.

## 3.2 Neighborhood

Most packages do not impose any restrictions on the neighborhood used. Often a coordinate-like notation for referencing relative neighbors is available and the neighborhood is given implicitly by the collection of all relative neighbors referenced. At least CDL allows the (relative) coordinates of neighbors to be computed by the local transition rule and to be stored in a register. This may simplify the formulation of some algorithms but makes the determination of the neighborhood more difficult.

The Margolus neighborhood is considered in subsection 3.5 below.

## 3.3 Set of States

Size.    Several systems restrict the size of the set of states to 256. One is then sometimes even urged to use the integers from $[0 \ldots 255]$ as states (which can be given names) and has to fiddle with the bits oneself. DDLab only allows one bit per cell. This has to do with the special kind of explorations it is intended for. All other systems (including CANL, CARPET, CDL, Cellang and Hical) allow an arbitrary number of states.

Structure. If the state becomes too large it obviously becomes important to be able to speak about its structure and to have a good notation. Often the memory of a cell can be subdivided into *registers* which can store values of different data types. This includes numerical as well as enumeration types. Hical allows to distinguish between integers and natural numbers (and to specify the exact number of bits to be used). Cellang and CDL allow integer subtypes of arbitrary ranges. CANL, CARPET, CDL and Hical offer at least one floating point type. Cellang and CDL and also offer the possibility to speak about arrays of values.

## 3.4 Transition Function

In Carpet, CamSim, CANL, CaSim, CAT, CDL, cellsim, Cellang, Ceprol, and Hical the local transition rule is specified in imperative languages offering (more or less) the usual operators for building expressions and the usual control structures. Furthermore many of them allow the definition of (auxiliary) functions which may be called.

In Fundef, SDL and some other languages the local transition rule is essentially specified as a set of rules. These may contain "don't cares" in the place of cells which have no influence on the resulting state in the current local configuration.

CDM offers the possibility to update cells asynchronously.

Most languages allowing the formulation of probabilistic rules do this via calls to a "random" function (e.g. CANL, CARPET, CDL, Cellang). Hical offers probabilistic statements which essentially look like case statements with probabilities assigned to the different branches. Of course there is also always the possibility the generate simple pseudo random bits by a (sub-)CA.

## 3.5 Further Extensions

Partitioned CA. They can be formulated naturally in any language where cells can be divided into registers.

Block Rules. We know of no system offering the possibility to specify block rules as an intrinsic feature. In CamSim and in Hical one can exploit other features of the languages to get something which looks very much like specifying a block rule.

Moving Entities. CDL and Cellang offer (different) concepts of moving entities (called mobs and agents respectively) as an intrinsic feature.

Temporal Inhomogeneity. On the conceptual level temporal inhomogeneity is easily achieved by providing a *global variable* containing the number of steps already simulated, which can be read (but not written) in the program for the local transition rule. This is possible in CARPET and Cellang (at least). On the other hand temporal inhomogeneity may complicate the code generation for special hardware simulators.

Of course simple version of temporal inhomogeneity (e.g. cyclically changing the behavior of all cells) can be realized by adding a register to the cells which is used as a modulo counter.

Spatial Inhomogeneity.   In CDL the local transition rule of a cell can test whether it is positioned at the border of the grid and where. CARPET goes even further and allows the rule to use the Cartesian coordinates of the cell. Hical allows smaller modules to be put together on the (possibly larger) lattice.

Again simple versions of spatial inhomogeneity can be achieved by extending the set of states by registers initialized with "spatial information".

Hierarchical Aspects.   These are present in CANL and Hical. The main difference is that in CANL the number of transition steps may vary widely on different levels on the hierarchy while in Hical this is not possible.

## 3.6   Compilation Issues

While in a few cases the simulation system is an interpreter, most SP contain a compiler which translates the CA program into a program in a general purpose (high or low level) language, e.g. C. The efficiency of the resulting code depends on several aspects. One is the semantics of the CA language which should not promise too much. A second important aspect is the presence (or absence) of a well chosen syntax. There exist very simple examples of transition functions which can be compiled into very efficient code, but only if certain constructs are syntactically easy to discover as such.

A third aspect which is difficult on a more general level is the choice of appropriate coding and storage strategies [21]. Assuming memory words have $w$ bits and $b < w$ bits are needed for storing the state of one state, one can either store one state in one word, or store $\lfloor w/b \rfloor$ states in one word (using the words as microvectors [2] as the Cellang compiler does), or store $w$ states in $b$ words by storing the $b$ bits of each state at the same bit position of $b$ words (multi-spin coding). It usually depends on the local transition rule which alternative allows the fastest simulation.

# Acknowledgments

The author gratefully acknowledges interesting discussions with Andreas Beckers, Claudia Di Napoli, Christian Hochberger and Martin Kutrib.

# References

[1] E. D. Adamides, Ph. Tsalides, and A. Thanailakis. Hierarchical cellular automata structures. *Parallel Computing*, 18:517–524, 1992.

[2] Micah Beck and Antonio Castellanos. Vector processing on scalar architectures. Technical Report CS-94-247, Computer Science Department, Univ. of Tennessee, Knoxville, 1994.

[3] Andreas Beckers. *Ein Ansatz zur hierarchischen Programmierung von Zellularautomaten.* Diploma thesis, Fakultät für Informatik, Universität Karlsruhe, 1995.

[4] Claudia Di Napoli, Maurizio Giordano, Mario Mango Furnari, Franco Mele, and Renata Napolitano. CANL: a language for cellular automata network modeling. In Vollmar et al. (eds.) *Parcella '96*, nr. 96 in Mathematical research, pp. 101–111, Berlin, 1996. Akademie Verlag.

[5] Andreas Döring. *Optimierung und Strukturierung von Zellularräumen.* Diploma thesis, Fakultät für Informatik, Universität Karlsruhe, 1995.

[6] Jérôme Olivier Durand-Lose. Partitioning automata, cellular automata, simulation and reversibility. Research Report 95-01, Ecole Normale Supérieure de Lyon, Lyon, 1995.

[7] J. Dana Eckart. A cellular automata simulation system: Version 2.0. *SIGPLAN Notices,* 27:99–112, 1992.

[8] Ulrich Golze. (A-)synchronous (non-)deterministic cell spaces simulating each other. *Journal of Computer and System Sciences,* 17:176–193, 1978.

[9] Armin Hemmerling. Concentration of multidimensional tape-bounded systems of Turing automata and cellular spaces. In L. Budach (ed.), *FCT '79,* pp. 167–174, Berlin, 1979. Akademie-Verlag.

[10] David Hiebeler. A brief review of CA packages. *Physica D,* 45:463–476, 1990.

[11] Christian Hochberger and Rolf Hoffmann. CDL — a language for cellular processing. In *MPCS '96,* pp. 41–47, Ischia, 1996.

[12] Robert Scott Ladd. *C++ Simulations and Cellular Automata.* M&T Books, New York, NY, 1995.

[13] Meena Mahajan and Kamala Krithivasan. Some results on time-varying and relativised cellular automata. *International Journal of Computer Mathematics,* 43(1/2):21, 1992.

[14] Kenichi Morita. Computation-universality of one-dimensional one-way reversible cellular automata. *Information Processing Letters,* 42:325–329, 1992.

[15] Jürgen Ruf. *Untersuchung von erweiterten Zellularautomaten mit variabler Nachbarschaft.* Diploma thesis, Fakultät für Informatik, Universität Karlsruhe, 1996.

[16] Friedhelm Seutter. CEPROL: A cellular programming language. *Parallel Computing,* 2(4):327–333, December 1985.

[17] Hans B. Sieburg and Oliver K. Clay. The cellular device machine development system for modeling biology on the computer. *Complex Systems,* 5:575–601, 1991.

[18] Giandomenico Spezzano and Domenico Talia. CARPET: A programming language for parallel cellular processing. In *Proc. 2nd European School on Parallel Programming Environments (ESPPE '96),* pp. 71–74, Alpe d'Huez, April 1996.

[19] Tommaso Toffoli and Norman H. Margolus. *Cellular Automata Machines – a New Environment for Modeling.* MIT Press, Cambridge, MA, 1986.

[20] John von Neumann. *Theory of Self-Reproducing Automata.* University of Illinois Press, 1966. edited and completed by Arthur W. Burks.

[21] Jörg Richard Weimar. Lecture notes: Simulation with cellular automata. http://www.tu-bs.de/institute/WiR/weimar/, 1996.

[22] Juraj Wiedermann. Parallel Turing machines. Technical Report RUU-CS-84-11, University Utrecht, Utrecht, 1984.

[23] Thomas Worsch. Programming environments for cellular automata. Technical report 37/96, Universität Karlsruhe, Fakultät für Informatik, November 1996.

# A Lattice Boltzmann Wave Model and its Applications

Bastien Chopard and Pascal O. Luthi
CUI, University of Geneva

## Abstract

We present a lattice Boltzmann model for simulating wave propagation in complex environments. We illustrate the behavior of our model on several applications and discuss its ability to describe solid body motion and fracture phenomena.

## 1   The Lattice Boltzmann Approach

Lattice Boltzmann (LB) models are a powerful approach to simulate on a computer many physical phenomena and, in particular fluid flows[1]. These models describe the physical quantities of interest in terms of fields $f_i(\vec{r}, t)$ defined for each point $\vec{r}$ of a lattice and each discrete time step $t = n\tau$. The index $i$ is associated with the lattice directions. For instance, in a 2D square lattice, $i$ runs from 1 to 4 and labels direction left, up, right and down, respectively.

In a hydrodynamical application, $f_i(\vec{r}, t)$ represents the average number of particles entering site $\vec{r}$ at time $t$ with velocity $\vec{v}_i$. The velocities $\vec{v}_i$ corresponds to a motion along lattice direction $i$ so that, in one time step $\tau$, one lattice spacing $\lambda$ is traveled. A zero velocity $\vec{v}_0 = 0$ can be introduced too, in order to describe a population $f_0$ of rest particles.

Macroscopic quantities like the density $\Psi$ or the momentum $\vec{J}$ can be defined using the standard procedure of statistical mechanics[2], namely

$$\Psi = \sum_i f_i \qquad \vec{J} = \sum_i f_i \vec{v}_i$$

A Boltzmann equation[2] is a balance equation which describe how the populations $f_i$ are re-distributed after an interaction. A general lattice Boltzmann equation reads

$$f_i(\vec{r} + \tau \vec{v}_i, t + \tau) - f_i(\vec{r}, t) = \Omega_i(f)$$

where $\Omega(f)$ is called the collision term. It is usually a nonlinear function of the $f_i(\vec{r}, t)$'s.

When the dynamics is based on specific "microscopic" interactions (as is the case with lattice gas automaton models[3]), an explicit expression for $\Omega$ can be obtained. However, the collision term can be expressed in a more abstract way too: in the so-called LBGK models[4], the dynamics of the functions $f_i$ is governed by a LB equation where the interaction is simply given by a relaxation term

$$f_i(\vec{r} + \tau \vec{v}_i, t + \tau) - f_i(\vec{r}, t) = \frac{1}{t_{rel}} \left( f_i^{(0)}(\vec{r}, t) - f_i(\vec{r}, t) \right) \qquad (1)$$

where $f_i^{(0)}(\vec{r}, t)$ is the so-called local equilibrium distribution and $t_{rel}$ the relaxation time. The function $f_i^{(0)}$ is the key ingredient of a LBGK model since it actually contains the properties of the physical process which is studied: this is the distribution to which the dynamics spontaneously relaxes and which is, therefore, intimately related to the nature of the system.

LBGK models for hydrodynamic flows assume a local equilibrium which is polynomial (quadratic in the local velocity field and linear in the local density of particles). The coefficient of each term is adjusted so that mass and momentum are conserved and the Navier-Stokes equation reproduced.

## 2 Wave Equation

In this paper we would like to use the LB formalism to simulate wave propagation in complex media. Sound waves are obtained in hydrodynamics by neglecting nonlinear velocity terms in the Navier-Stokes equation. Thus, in order to devise a wave model, we simply impose that the local equilibrium function is linear in the conserved quantities $\Psi$ and $\vec{J}$

$$f_i^{(0)} = a\Psi + b\frac{\vec{v}_i \cdot \vec{J}}{2v^2} \qquad \text{if } i \neq 0, \text{ and} \qquad f_0^{(0)} = a_0\Psi \qquad (2)$$

The parameters $a$, $b$ and $a_0$ are chosen so that conservation of $\Psi$ and $\vec{J}$ are satisfied, that is

$$\sum_i (f_i(\vec{r} + \tau \vec{v}_i, t + \tau) - f_i(\vec{r}, t)) = 0$$

$$\sum_i \vec{v}_i (f_i(\vec{r} + \tau \vec{v}_i, t + \tau) - f_i(\vec{r}, t)) = 0$$

From (1), this amounts to asking that

$$\Psi = \sum_i f_i^{(0)} \qquad \vec{J} = \sum_i \vec{v}_i f_i^{(0)}$$

For a two-dimensional square lattice, we have $\sum_{i=1}^4 \vec{v}_i = 0$ and $\sum_{i=1}^4 v_{i\alpha} v_{i\beta} = 2v^2 \delta_{\alpha\beta}$, where $\alpha$ and $\beta$ label the spatial coordinates. Thus, the conservation of $\Psi$ and $\vec{J}$ yields the condition

$$a_0 + 4a = 1 \qquad b = 1$$

The evolution of $\Psi$ can be computed using a multiscale Chapman-Enskog expansion[3, 5, 6]. To this end, equation (1) can be solved by writing $f_i = f_i^{(0)} + \epsilon f_i^{(1)} + \epsilon^2 f_i^{(2)} + ...$, where $\epsilon$ is a small parameter which reflects the fact that the macroscopic observation time and length scales are much larger than the lattice spacing and the time step. Then, the finite difference $f_i(\vec{r} + \tau \vec{v}_i, t + \tau) - f_i(\vec{r}, t)$ is expressed in terms of space and time derivatives, using a Taylor expansion up to second order in $\lambda$ and $\tau$. To describe the limit $\lambda \to 0$, $\tau \to 0$ is it important to specify how $\lambda$ and $\tau$ both go to zero. It is expected *a priori* that two time scales should be considered: the first accounts for the convective effects and the second describes diffusive effects. Thus, we write the derivatives as $\partial_t = \epsilon \partial_{t_1} + \epsilon^2 \partial_{t_2}$ and $\partial_\alpha = \epsilon \partial_{\vec{r}'_\alpha}$, where $t_1$, $t_2$ and $\vec{r}'$ are the new time and space variables.

After straightforward algebra (where terms of the same order in $\epsilon$ are collected), the functions $f_i^{(1)}$ can be computed and we obtain that equation (1) implies

$$\partial_t \Psi + \partial_\beta J_\beta = 0 \quad (3)$$

$$\partial_t J_\alpha + 2av^2 \partial_\alpha \Psi + (2t_{rel} - 1) \left[ a\tau v^2 \partial_\alpha \mathrm{div} \vec{J} - \frac{\tau}{4v^2} T_{\alpha\beta\gamma\delta} \partial_\beta \partial_\gamma J_\delta \right] = 0 \quad (4)$$

where $T_{\alpha\beta\gamma\delta} = \sum_i v_{i\alpha} v_{i\beta} v_{i\gamma} v_{i\delta}$. In these equations and the rest of the discussion, summation over repeated greek indices is assumed.

The tensor $T_{\alpha\beta\gamma\delta}$ is not isotropic for a 2D square lattice (it depends on a particular orientation of the lattice). By choosing $t_{rel} = 1/2$ this anisotropy is removed. For fluid models, $t_{rel} = 1/2$ corresponds to having a zero viscosity. Here, also, this choice suppresses dissipation and makes the evolution rule (1) invariant under time reversal (which is indeed a very fundamental property of the wave equation). Note that in fluid models, the limit $t_{rel} = 1/2$ is numerically unstable but here, it is not the case because the dynamics is linear in the $f_i$'s.

With $t_{rel} = 1/2$ equation (4) becomes

$$\partial_t J_\alpha + 2av^2 \partial_\alpha \Psi = 0 \quad (5)$$

Equations (5) and (3) can be combined to give

$$\partial_t^2 \Psi - 2av^2 \nabla^2 \Psi = 0$$

which is a wave equation with propagation speed $c = v\sqrt{2a}$ (where $v = \lambda/\tau$ is the speed at which information travels).

A very interesting feature of this wave model is that the propagation speed $c$ can be adjusted from place to place by choosing the spatial dependency of $a$. Provided that $a_0 + 4a = 1$ and $a_0 \geq 0$, the larger possible value is $a = 1/4$ and corresponds to a maximum velocity $c_0 = v/\sqrt{2}$. Therefore media with different refraction indices

$$n = \frac{c_0}{c} = \frac{1}{2\sqrt{a}}$$

can be modeled.

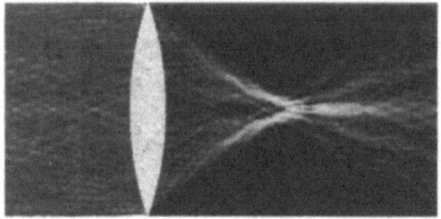 

Figure 1: Simulation with the LB wave model: focusing of light by a convex lens where the propagation speed is smaller than in vacuum (left). focusing by a parabolic mirror (right).

# 3   The Lattice Boltzmann Wave Model

Relation (2) for $f_i^{(0)}$ can be substituted into equation (1). Using that $t_{rel} = 1/2$, $a = 1/(4n^2)$, $a_0 = (n^2 - 1)/n^2$, we get the following simple evolution rule for the LB wave model

$$
\begin{aligned}
f_i(\vec{r} + \tau\vec{v}_i, t + \tau) &= \frac{1}{2n^2}\Psi - f_{i+2}(\vec{r}, t) \qquad \text{if } i \neq 0 \\
f_0(\vec{r}, t + \tau) &= 2\frac{n^2 - 1}{n^2}\Psi - f_0(\vec{r}, t)
\end{aligned}
$$

$$(6)$$

where $n > 1$ is the space dependent refraction index and $\Psi = \sum_{j=0}^{4} f_i(\vec{r}, t)$. Here, $i + 2$ denotes the direction opposite to $i$. Equation (6) describes free propagation. When $n = 1$ and the initial distribution of $f_0$ is zero, only $f_1$, $f_2$, $f_3$ and $f_4$ are necessary in the dynamics.

Note that the above numerical scheme has already been derived by other authors, but in a different context or with a different approach[7, 8, 9]. The present approach is more compact makes it easy to generalize the model in higher dimension or to another lattice.

Figure 1 (left) shows a simulation of equation (6) in a situation where two media are present. A plane wave is produced in medium $M_1$ by forcing a sine oscillation for the $f_i$'s on some vertical line. The wave propagates at speed $c_0$ till it penetrates in medium $M_2$ with has the shape of a convex lens. There, propagation speed is set to $c < c_0$. The shape of the lens naturally produces a focusing of the energy when the wave re-enters medium $M_1$.

A natural interpretation of our LB wave model is to assume that the $f_i$'s represent some physical fields (a local deformation or deviation from an equilibrium state). These fields propagate on the lattice and are scattered when reaching a site. Figure 2 illustrates this picture for a situation where $n = 1$ and $f_0 = 0$. Reflection on obstacles are then easy to defined: we may assume that an incoming flux $f_i$ will bounce back when a lattice site is not permeable. Since a reflecting wave change sign when it reaches a perfectly absorbing obstacle, we shall model pure reflection as

$$
f_i(\vec{r} + \tau\vec{v}_i, t + \tau) = -f_{i+2}(\vec{r}, t) \qquad (i \neq 0) \tag{7}
$$

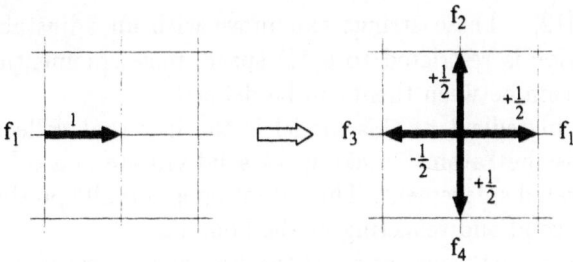

Figure 2: Scattering of an incoming flux $f_1 = 1$ at a lattice site, according to equation (6). The value of each outgoing flux is indicated on the arrows.

An example of a wave reflected on a parabolic mirror is shown in figure 1 (right). Each lattice site in the black region is a perfect reflector obeying equation (7). As a result of the collective effect of these mirror sites, we observe that the incoming plane wave concentrates at the focal point of the parabola.

An absorbing media can also be simulated. In this case we will assume that the oscillating field $\Psi$ is not conserved but decreases. To obtain this effect we multiply $a_0$ and $a$ in equation (2) with an attenuation factor $0 < \mu < 1$. With $f_0^{(0)} = \mu a_0 \Psi$ and $f_i^{(0)} = \mu a \Psi + (\vec{v}_i \cdot \vec{J})/(2v^2)$, we have

$$\sum_{i=0}^{4} (f_i(\vec{r} + \tau \vec{v}_i, t + \tau) - f_i(\vec{r}, t)) = -(1 - \mu)\Psi$$

and the evolution reads

$$f_i(\vec{r} + \tau \vec{v}_i, t + \tau) = \frac{\mu}{2n^2}\Psi - f_{i+2}(\vec{r}, t) \qquad f_0(\vec{r}, t + \tau) = 2\mu\frac{n^2 - 1}{n^2}\Psi - f_0(\vec{r}, t)$$

From (6) and (7) we see that absorption is equivalent to mix free propagation with probability $\mu$ and perfect reflection with probability $1 - \mu$.

This LB wave model has been successfully used to predict wave propagation in an urban environment where the buildings reflect, scatter and absorb energy in a complicated way[10, 11]. This application is important for the development of the future mobile phone networks. In the remaining of this paper we would like to show how our LB wave model can also be used to represent solid body motion and fracture phenomena.

# 4   Modeling Solid Body

Whereas cellular automata and lattice Boltzmann methods have been largely used to simulate systems of point particles which interact locally, modeling a solid body with this approach (i.e modeling an object made of many particles that maintains its shape and coherence over distances much larger than the interparticle spacing) has remained mostly unexplored. A successful attempt to model a *one-dimensional* solid (also termed a string) as a cellular automata is described in[12]. These strings can move with an adjustable speed and, when their motion is restricted to a 1D space, mass, momentum and energy conserving collision between them can be defined.

The crucial ingredient of this model is the fact that collective motion is achieved because the "atoms" making the solid vibrate in a coherent way and produce an overall displacement. This vibration is actually produced by a wave traveling in the solid and reflecting at the boundary.

Thus a microscopic wave model is the first step to simulate a solid body. The main difficulty when extending the 1D cellular automata model to 2D objects is that particles moves off lattice (a 2D wave spread in all direction and cannot be described with integer arithmetics).

However, we shall see that the LB wave model described in the previous section is the natural way to extend the 1D cellular automata string dynamics. The distance separating consecutive particles turns to obey the dynamics expressed by equation'(6.

## 4.1 The 1D Model

We start the discussion by briefly recalling the rule of the cellular automata model for 1D solid bodies (string) described in [12].

A string can be thought of as a chain of masses linked by springs. More precisely, it is composed of two kinds of particles, say the white ones and the black ones, which alternate along a line. Two successive particles along the chain are either nearest neighbors or separated by one empty cell. Figure 3 shows such a chain composed of five particles. The time evolution has two phases. First,

$$t=0 \qquad\qquad t=1 \qquad\qquad t=2$$

Figure 3: A string with 5 particles and its motion at subsequent time steps.

the black particles are held fixed and the white ones move according to the rule explained below. Then, the white particles are held fixed and the black ones move with the same prescription.

The rule of motion is the following: let us consider the case where the white particles moves. The new configuration of the string is obtained by interchanging the spacings that separate the white particles from their two adjacent black neighbors at rest.

In other words, the motion of a white particle consists of a reflection with respect to the center of mass of its two black neighbor particles. Of course this rule is only valid for a particle inside the string and having exactly two neighbors. For the end particles, we assume that the motion is governed by the same dynamics provided that a virtual particle is added at the extremity of the string. This virtual particle is placed three sites away from the next to the last string particle. This is equivalent to ask that the last two particles are connected with a spring of length $r_0 = 3/2$.

Figure 3 illustrates the behavior of this rule (black particles motion followed by white particles motion).

## 4.2  The 2D Model

The above 1D model can be generalized as follows. Let us assume we have a two-dimensional lattice of black and white particles organized in a checkerboard fashion. Figure 4 shows a basic cell of such an "atomic" structure, namely a black particle surrounded by four white particles. In what follows, we shall only consider a four-field model, disregarding the rest field $f_0$.

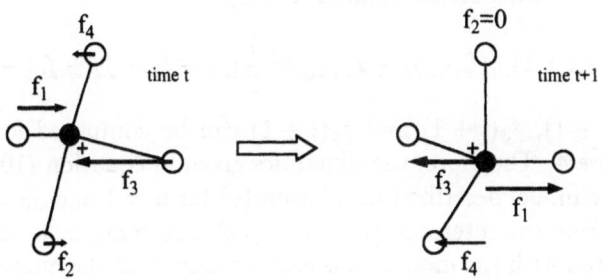

Figure 4: Alternate motion of black and white particles in a two-dimensional system. The quantities $f_1$, $f_2$, $f_3$, and $f_4$ measure the distance, along the x-axis, separating the central particle from its four connected neighbors. The cross indicates the location of the center of mass of the white particles. Other quantities $g_i$'s can be defined to describe the separation along the vertical y-axis.

As in the one-dimensional case, the black particles move at even time step and the white ones at odd time steps. In figure 4, the motion is done relative to the positions of the four connected white neighbors at rest. The rule of motion for the black particles is to jump to the symmetrical position with respect to the center of mass (the cross in the figure) of the four white surrounding particles.

It is important to note that, in this model, the coupling between adjacent particles is not given by the Euclidean distance but decouples along each coordinate axis. This breaks the rotational invariance but simplifies very much the model.

Let us denote the location of the black particle by $\vec{r}_{i,j} = (x_{i,j}, y_{i,j})$. The surrounding white particles will be at positions $\vec{r}_{i-1,j}$, $\vec{r}_{i+1,j}$, $\vec{r}_{i,j-1}$ and $\vec{r}_{i,j+1}$. We define their x-axis separation to the central black particle as

$$f_1(i,j,t) = x_{i-1,j}(t) - x_{i,j}(t) \qquad f_3(i,j,t) = x_{i+1,j}(t) - x_{i,j}(t)$$
$$f_2(i,j,t) = x_{i,j-1}(t) - x_{i,j}(t) \qquad f_4(i,j,t) = x_{i,j+1}(t) - x_{i,j}(t)$$
$$(8)$$

Similarly, we can define $g_1$, $g_2$, $g_3$, and $g_4$ as the distance, along the y-axis, separating the white particles from the black ones.

If the origin of the coordinate system is the black particle (that is $\vec{r}_{i,j}(t) = 0$), the x-coordinate $X_{CM}$ of the center of mass of the four white particles is $X_{CM} = (f_1 + f_2 + f_3 + f_4)/4$. Because the dynamics is to move to the symmetrical position with respect to the center of mass, the new position of the black particle is then

$$x_{i,j}(t+1) = 2X_{CM} = \frac{1}{2}[f_1 + f_2 + f_3 + f_4] \qquad (9)$$

The motion of the black particle results in a new distribution of the $f_i$'s. Since at time $t+1$ the white particles will move and the black will be stationary, the $f_i(t+1)$'s have now to be computed respective to each white particle location. For instance, $f_1(i+1, j, t+1)$ will be the separation to the black particle as seen by the particle located at $\vec{r}_{i+1,j}$

$$f_1(i+1, j, t+1) = 2X_{CM} - x_{i+1,j} = \frac{1}{2}[f_1 + f_2 + f_3 + f_4] - f_3 \qquad (10)$$

Similarly, $f_2(t+1)$, $f_3(t+1)$ and $f_4(t+1)$ can be computed for each white particle in figure 4. Therefore, the dynamics given in equation (10) is identical to the LB wave model described in relation (6) for $n = 1$ and $f_0 = 0$.

Note that from the interpretation of a local deformation, we can define an energy associated with the model. It is easy to show that the quantity $\sum_{k=1}^{4} f_k^2$ is conserved by the above evolution rule. There is only potential energy here because from one time step to the next, the particles move from one extrema to the other. Thus they have no velocity when one look at them.

# 5 Solid Body Motion

Our aim is now to simulate the motion of a solid body using the above approach. A solid object is made of a checkerboard structure of black and white particles located at position $\vec{r}_{ij}$. The departure from the equilibrium position obeys the LB wave model and is described by the fields $f_k$ and $g_k$. The actual location of the particles in space can be reconstructed from (8) and a similar relation for $y_{ij}$ and $g_k$.

It necessary to define the rule of motion for the particles sitting at the boundary of the solid. We use the same strategy as explained in the 1D case, namely to add virtual particles at the extremities (see figure 5). These particles are positioned at a distance $2r_0$ where $r_0$ is the equilibrium distance between "atoms." Note that $r_0$ is different for the horizontal and vertical coupling (when the x-motion is considered, $r_0$ is zero for the vertical links but not for the horizontal ones).

Figure 6 shows a simulation of a two-dimensional LB solid body. An initial momentum is given to the object by choosing non zero the values of $f_k$ and $g_k$ across the system. The propagation of this initial deformation produces an overall motion. The center of mass of the object follows the straight line shown in the figure until the object boundary reaches the limit of the simulation space, which acts as a reflecting wall. By forbidding the particles to cross this wall, the entire object naturally bounces back. During such a collision, the solid deforms but keeps its integrity.

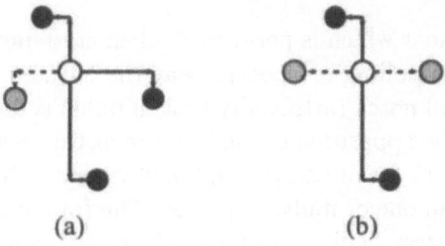

(a)                                        (b)

Figure 5: Virtual (gray) particles are added for each missing bond. In (a) a particle is missing on the left of the middle white particle; a virtual particle is placed at a distance $2r_0$ from the rightmost black particle and at the same height. In (b) both left and right bonds are absent; two virtual particles are added, at locations $\pm r_0$ from the white particle and at the same height.

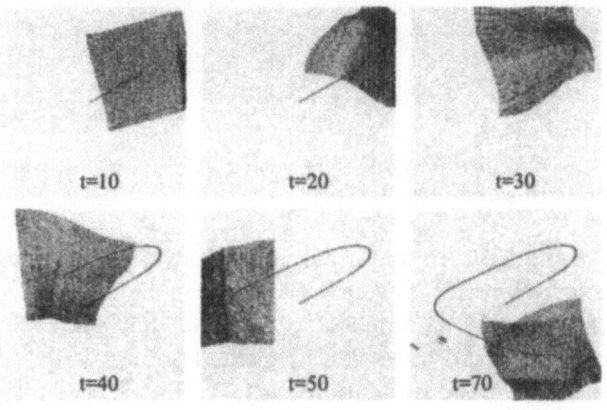

Figure 6: Motion of a deformable object made of a square structure of particles interacting according to the lattice Boltzmann wave rule. The solid bounces back on the boundaries of the simulation.

# 6  Fractures

Another important application of our LB solid body model is the study of a fracture process. How things breaks is still an important problem in science for which one lacks theory and no satisfactory understanding is yet achieved[13].

The key idea when using our approach as a model of dynamic crack is to assume that a bond linking two connected atoms may break if the local deformation exceeds some given threshold. This threshold can possibly be different for each bond and spatial disorder can be introduced in this way. Once a bond is broken, the atoms on each side of the crack behave as free ends and their dynamics is governed by the presence of virtual particles, as explained previously (figure 5). A broken link weakens the material because a local deformation can no longer be distributed uniformly among the four neighbors. Usually, the next bond to break is nearest neighbor of an already broken bond.

A typical experiment which is performed when studying fracture formation is to apply a stress by pulling in opposite way the left and right extremities of a solid sample. A snall notch (artificially broken links) is made in the middle of the sample to favor the apparition of the fracture at this position. Once a given strain is reached, a crack forms and propagate from that notch through the bulk, breaking the system in one or multiple pieces. The fracture is perpendicular to the direction of the stress. This situation is illustrated in figure 7, which shows a simulation obtained with our LB solid model. Each dot in the figure shows the position of an atom.

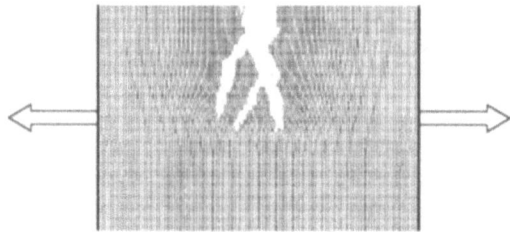

Figure 7: Fracture obtained in a LB solid with $128 \times 128$ atoms when applying an opposite force on both sides of the sample.

The shape of the fracture we obtain is qualitatively similar to what is observed in real experiment[13]. Several situations can be reproduced, depending on the value of the model parameters. It is found that adding some attenuation in the motion (see end of section 3) yields fractures with less branching. Figure 8 shows some of the simulation outputs. In figure (b) no damping of the wave is included while, in (a) a damping factor $\mu = 0.92$ is added. Figures (c) and (d) have less disorder than (a) and (b) in the sense that the threshold varies spatially only a little. The damping in (d) is $\mu = 0.91$, slightly stronger than in (c) ($\mu = 0.92$). The stretching rate (i.e the displacement of the solid boundary at each time step) is the same for all experiments. In the above simulations, once the fracture starts propagating, the external stress is turned off.

We have measured the propagation speed of the fracture by recording the location of the crack tip $l(t)$ for each time step. In case of branching we consider the most advanced crack. Figure 9 shows the average velocity $v(t) = l(t)/t$ of the propagation fracture as a function of time. These measurements made from our simulation are in qualitative agreement with experimental data. In particular the crack speed is slower than the speed of sound (which is here $c_0 = 1/\sqrt{2}$ in lattice units) and it is faster when the fracture is complex.

# References

[1] Y.H. Qian, S. Succi, and S.A. Orszag. Recent advances in lattice boltzmann computing. In D. Stauffer, editor, *Annual Reviews of Computational*

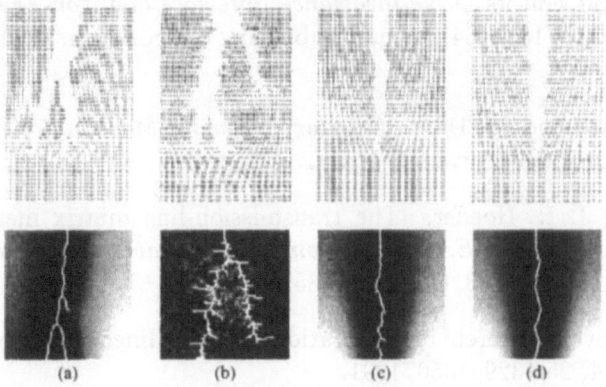

Figure 8: Fracture (top) and the corresponding map of the broken bond (bottom) for several runs with different parameters.

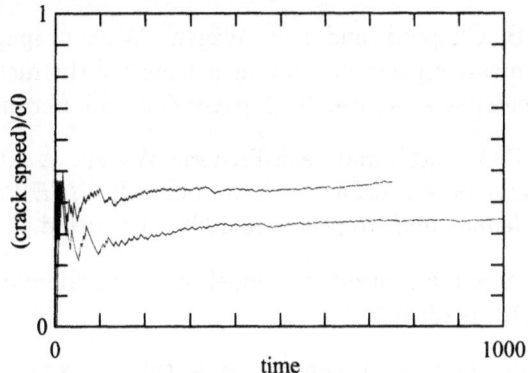

Figure 9: Crack propagation speed measured in the LB fracture simulation. The upper and lower curves correspond to the fractures shown in figure 8 (b) and (d), respectively.

*Physics III*, pages 195–242. World Scientific, 1996.

[2] P. Résibois and M. Leener. *Classical Kinetic Theory of Fluids*. John Wiley, 1977.

[3] U. Frisch, D. d'Humières, B. Hasslacher, P. Lallemand, Y. Pomeau, and J.-P. Rivet. Lattice gas hydrodynamics in two and three dimension. *Complex Systems*, 1:649–707, 1987. Reprinted in Lattice Gas Methods for Partial Differential Equations, Ed. G.Doolen, p.77, Addison-Wesley 1990.

[4] Y. H. Quian, D. d'Humières, and P. Lallemand. Lattice bgk models for navier-stokes equation. *Europhys. Lett*, 17(6):470–84, 1992.

[5] B. Chopard. Cellular automata modeling of hydrodynamics and reaction-diffusion processes: Basic theory. In A. McKane, M.Droz, J.Vannimenus,

and D.Wolf, editors, *Scale invariance, Interface and Non-Equilibrium Dynamics*, pages 133–164. Plenum Publishing Corporation, 1995. NATO ASI Series.

[6] B. Chopard and M. Droz. *Cellular Automata Modeling of Physical Systems*. Cambridge University Press. to appear.

[7] Wolfgang J. R. Hoeffer. The transmission-line matrix method. theory and applications. *IEEE Transaction on microwave theory and techniques*, MTT-33(10):882–893, October 1985.

[8] H. J. Hrgovcić. Discrete representation of the $n$-dimensional wave equation. *J. Phys. A*, 25:1329–1350, 1991.

[9] C. Vanneste, P. Sebbah, and D. Sornette. A wave automation for time-dependent wave propagation in random media. *Europhys. Lett.*, 17:715, 1992.

[10] P. O. Luthi, B. Chopard, and J.-F. Wagen. Wave propagation in urban microcells: a massively parallel approach using the tlm method. In *Lecture notes in computer science; vol. 1041*, pages 429–435. Berlin Springer, 1996.

[11] B. Chopard, P.O. Luthi, and Jean-Frédéric Wagen. A lattice boltzmann method for wave propagation in urban microcells. *IEE Proceedings - Microwaves, Antennas and Propagation*, 1996. submitted.

[12] B. Chopard. A cellular automata model of large scale moving objects. *J. Phys.*, A(23):1671–1687, 1990.

[13] Michael Marder and Jay Fineberg. How things break. *Physics Today*, pages 24–29, September 1996.

# Complex Systems and Core Competencies for the Industrial Innovation

Franco Forlani

AIRI (Italian Association for Industrial Research)

Rome, Italy

## 1.    Introduction

As stressed by the organizers of the present Conference in the Program they proposed, cellular automata are more and more taken into consideration for attaching problems of interest for the industry. In these days, in fact, the opportunity to become acquainted with competences on cellular automata is under evaluation by a certain number of industrial Companies. However, also in case of favourable decision, many of them have not yet decided in which way to accomplish the acquaintance of such competencies. As in many other instances, that can be done in one or in the other of the two following ways:

1) by establishing a team in the R&D Section, initially in charge of sucking everything available as state-of-the-art in that field and, then, of attaching problems and of implementing matters of interest to the Company;

2) by "outsourcing", i.e. by perfectionating agreements with Institutions external to the Company and trusted to hold an up-dated know-how in that field; the Company commits the matters of its interest in that field to such Institutions' care.

In general, to apply to outsourcing is frequently considered in order not to loose a close connection with the world-wide technological evolutions and innovations, but, at the same time, to minimise investments with long-term return.

Aim of the present paper is to point out elements to give an answer to the following question: when does a Company choose to invest directly on the development and on the maintenance of a competence and when does it prefer to outsource for it?

Finally, a discussion is proposed on the availability of human resources properly prepared to deal with advanced competencies either in industrial Companies or in academic Institutions, by stressing the particular situation of Italy on such a matter.

## 2.    Core competencies

To find possible answers to the above-posed question let me start with considering what "competence" means to an industrial Company. Many definitions have been proposed [1] [2] and all of them have in common a few concepts: competence is a harmonised bundle of skill, competence transcends any particular product or service offered by the Company, competence enhances competitiveness and prosperity of the Company. Moreover, it is commonly recognised that to get competencies, to maintain and renovate them and to manage them according to the strategic business plan is one of the most effective way for a Company to get success in its business.

Among the different types of competencies required by a Company the identification of that category defined as "core competencies" can be done by taking advantage of a similitude proposed by Prahalad and Hamel [3]. Core competencies are similar to the roots of a tree, whose trunk symbolises the "core business" of the Company. The branches of the tree represent the "core technologies" and the fruits the "end products". By going ahead with such a similitude, there exist a cross-fertilisation between the end products-fruits and the core competencies-roots, just as in the natural cycle. Some fruits fall down, make fertile the soil and provide nutrition to the roots. New lymph permeates the trunk and arriving through the branches to the fruits make them more plentiful and tasty. Similarly, end products facing competition on the market stimulate fertilisation and nutrition to the core competencies and make the end products improved and the competitivity on the market enhanced.

Going deeper in the analysis of core competencies, they can be divided in two groups. A first group includes those competencies more tightly connected to the skills and to the disciplines typical of the specific core technologies of the Company. They are similar in Companies operating in the same industrial sector, but different for different industrial sectors. For instance, in a Company manufacturing and selling microprocessor chips the technology-oriented core competencies appear to be different from those of a Company operating in oil and gas exploration and production.

The second group identifies non-technological core competencies. Even if, in general, they can have a certain degree of specificity, they can be essentially the same also in Companies operating in different industrial sector. Rather, they are more typical of the mission assigned or chosen by a Company. To make clearer this point, let us consider Companies having the mission to develop innovative processes or products. Examples of non-technological core competencies for them should be the technology acquisition and transfer, the knowledge up-dating or the vision of future scenarios, independently of the fact the industrial sector is computer or biotechnology or chemistry. On the other hand, such non-technological core competencies can be different for Companies having their business in commodity or in speciality market even if belonging to the same industrial sector.

Then, core competencies, even if immaterial, can be assimilated to the assets of a Company and, consistently, do not appear to be liable to outsourcing. That is particularly true for non-technological core competencies. As for the technological ones, some marginal skill could be chosen in the bundle and, if that is the case, outsourced. However, in order not to loose in harmonisation with the other skills a strong co-ordination has to be, in any case, exerted by the Company.

## 3. Other competencies

Of course, core competencies do not exhaust all the competencies a Company needs to carry on its own business. At least other two groups of competencies can be identified.

There are competencies never made explicit in listing the competencies of a Company, simply because they are quite expected, being essential for the operativity of a Company. To make an example, the accounts of employees' pay-roll, the compulsory contributions from the employer to the State, to the Assurances and so on and so forth are skills that all together make a competence, maybe not so rewarding as that required to prospect future scenarios, but nevertheless very necessary for the Company. We can call them "tacit competencies", just because even if acting they are commonly ignored.

The last group of competencies to speak about can be defined as "cropping-up competencies". They are associated to the fact that more and more frequently the extremely quick diffusion of information, - thanks, among others, to the spread use of informatics media -, makes some teams or single persons in the Company to become in contact with very advanced frontier knowledge. Some team or person may begin to try to use it for some particular application they are interested in. Through communication with other teams with different skills, through the testing of such knowledge in different problems and so on there is a new competence that crops up in the Company, even if not fully recognised as such. Probably, cellular automata provide an example of a frontier knowledge that in the last periods is just in that phase.

Both groups of competencies, tacit and cropping-up, are admittable to outsourcing. In effect, the tacit competencies are, essentially, made by the services necessary to a Company to operate. As well known, services are required to be effective, efficient and low cost, but as for competence they do not impact on the strategic choice of the Company about its process, product or innovation policy. If their outsourcing fulfils the above mentioned requirement, why not?

Perhaps, the situation is a bit more delicate when dealing with cropping-up competencies. The decider should evaluate whether the novelty is a skill that harmonised with others can prevail and become a competence or, more simply, it remains for the Company a skill helpful to be used when necessary; whether the new skill impact on a single process or product or it is pervasive, at least for a certain family of Company processes or products; whether the inner acquisition can provide competitivity advantages to the Company or not. In practice, the novelty should be checked face to face with the three attributes commonly recognised as qualifying a competence and reported at the beginning of § 2. If the check is positive, the Company should try to have the skill in house from soon. If not, maybe a policy of outsourcing and, at the same time, to get in house a capability of actively interfacing the external source should be preferable. In fact, in such a way the Company can proceed to a quick inner acquisition if the evolution makes it necessary, but is not committed in a full spending from soon. At present, cellular automata are probably seen in that way by several Companies.

## 4. Availability of human resources

The technological evolution requires new human resources, operating either in industrial Companies or in academic Institutions, dedicated to the novelties. In fact, a new horizon opened to the Science or to the Technology usually is complementary to and not a replacement of the existing knowledge.

A recent work has been done [4], and summarised in what follows, looking at how the number of researchers, industrial and academic all together, is changed in the different Countries in the last fifteen years. In order to compare the situation of one Country with another, Country by Country the number of researchers in a certain year has been correlated with the Gross National Product (GNP) of that year. All the Countries of the Western Europe plus USA and Japan have been examined.
It results that:

1) in all the Countries in the North of Western Europe and in USA the number of researchers per unit GNP is constant year by year, but the value of such a constant tends to decrease since 1980;

2) in Japan the number of researcher per unit GNP is systematically higher than the said constant;

3) in the Latin Countries of Europe, but especially in Italy, the number of researchers per unit GNP is systematically lower than the said constant.

Point 1) indicates that the Northern European Countries and USA act as a Company that, although increasing sales, does not proportionally innovate. That is typical of a Company strategy based on the penetration of emerging markets, for instance in the development Countries, with its traditional products. Point 2) means that, as well known, Japan bases its strategy on the technological innovation and that more than in the past is paying attention to the Science in addition to the Technology.

Point 3), among others, emphasises the anomalous situation of Italy. Extrapolating the data to year 2000, it results that at that time Italy is going to have a lack of 100.000 researchers in respect to the Northern European Countries, at least by using the above said parameter based on GNP. In such a situation it does not seem that the main problem for Italian industrial Company to be easy to follow sophisticated strategies of inner acquisition or outsourcing of skills and competencies. Unfortunately, Italy risks to have to outsource technological innovation *per se*.

# References

1.    Hamel J., Prahalad C.K., <u>Competing for the Future</u>, Harvard Business School, Boston Press

2.    Quinn J.B., Hilmer F.G., "Strategic Outsourcing", Sloan Management Review, Summer 1994, pp 43-55

3.    C. K. Prahalad, Hamel G., "La competenza distintiva delle aziende", Harvard Espansione <u>49</u>, Dec. 1990, p. 7

4.    F. Forlani, Sistemi & Impresa (in press)

# Contributed
# Papers

# Coupled Maps Lattice Dynamics on a Variable Space for the Study of Development

Franco A. Bignone*

Istituto Nazionale per la Ricerca sul Cancro, IST

L.go Rosanna Benzi, 10, I-16132 Genova, Italy

&

Dip. Biologia Animale e Genetica

Univ. di Firenze, Firenze, Italy.

e-mail: abignone@igecuniv.cisi.unige.it

### Abstract

The implementation of models to simulate genetic networks dynamics, in cell replication and development in Biological Systems, has to take into account the fact that interactions among genes and cells take place in a space of variable structure and size. In this paper we present modelling for the detailed reconstruction of cellular dynamics during early stages of development with Coupled Maps Lattices or Lindenmayer Grammars formalizations.

Keywords: Genetic Networks, Gene expression, Coupled map lattices, Cellular automata, Neural Networks, Caenorhabditis elegans, L-systems, Context sensitive grammars.

## 1 Introduction

Coupled maps lattices [CML], have been introduced in recent years as discrete models for spatially extended dynamical systems. In this context they have been used to study: spatially extended chemical reactions, interface dynamics, biological systems [1, 4], see reference [3] for a review. In all these cases the study has been limited to dynamics simulated, through parameter changes, on lattices of fixed size and shape from $t(0)$ onward. The lattice's morphologies have in general been defined over the ring or the torus with periodic boundary conditions. In some other cases the dynamics have been analyzed on particular lattice structures, such as the Sierpinski's Gasket, but always with a set of elements constant from the begininng of the simulations [2].

For biological systems, where cell replication, chemotactic movement, growth, and cell death occurs, this assumption is a limitation in respect of the main characteristics shown. Formalization of Biological Systems as lattices of fixed size is important to study basic dynamical properties. However for a more realistic study, if we use for example a CML formalization to study cell-cell interactions, it should be taken into account that local space-time interactions set lattice's size and shape – i.e. number of sites and topological distribution of signal sources – while biochemical signals travel through it driving growth, death and/or cell movement. This results in a continuous coupling between the studied dynamics and lattice structure. Biological phenomena that shows this kind of plasticity are known to be quite general and occur at early and late stages of development, in several pathologies – i.e. cancer progression – and during homeostasis [7, 10].

The different approaches introduced so far to study this problem can be subdivided into two main groups depending on the theoretical tools that have been used for the simulations. From the point of view of development, with a particular reference to plant morphologies, A. Lindenmayer and co-workers [6, 9, 11] have introduced the definition of L-systems, a particular type of formal grammars with parallel updating, for this pourpose.

Similar approaches, more classically oriented from the mathematical point of view, using PDE and geometric considerations, have also been developed in order to study the effect of cell-replication in the creation of structure.

In this work we introduce an approach that rely on a general definition, in terms of biochemical genetic networks, of the rules that set a context sensitive, generative, gammar. This model can be seen in turn as a general definition for a class of CMLs with variable degrees of freedom.

## 2  C. elegans as an L-system

The biological problem that we discuss here is the development of Caenoehabditis elegans [C.e.] up to the level of hatching. C.e. is a Nematode worm that has become one of the main experimental systems for the study of development and genetics.

A picture of the *four cells stage* of development is shown in Fig. 1. The egg surroundig the cells has an elliptical shape, and contain initially a single cell *the Zigote* that by subsequent subdivisions of its mass by cell replication give rise to 558 fully differentiated cells at hatching in the hermaphrodite, out of a total of 671 cells, 113 of which are eliminated by programmed cell death.

It has been known for a long time that several organisms, from Anellids to Gastheropods and Nematodes, have a fixed pattern of division and development. Detailed embryological studies have described in the past the pattern of replication in the embryo and genetic studies are on the way in order to define molecular determinants for such behaviour. In Fig. 2 is reported an example of the definitions currently used by embryologists, and the scheme for the first two divisions for the $AB$ lineage are reported. This results in a complex binary tree [12].

The development of this organism, from a descriptive point of view, it is thus very similar to an L-system as described by A. Lindenmayer [6].

L-systems has been conceived as a formal description of plant development, the emphasis was on plant topology. Subsequently several geometric interpretations of L-systems have been proposed, for example a turtle geometry interpretation of L-systems is used in [9]. The relationship between L-systems and formal grammars, such as Chomsky's grammars, lie in the method of applying productions. In Chomsky's grammars productions are applied sequentially, while in L-systems they are applied in parallel.

The definition of a formal language begins with the introduction of a finite nonvoid set of symbols, a finite alphabet, usually denoted by $V$. The elements

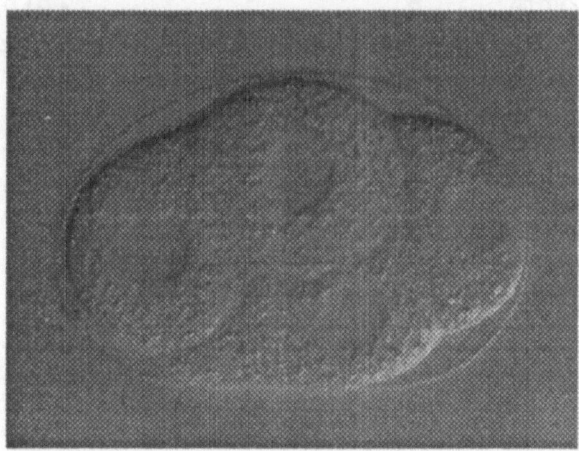

Figure 1: Caenorhabditis elegans, early stage of development, after the initial three divisions. Clockwise from the left side, cells are: $ABa$, $ABp$, $P_2$, $EMS$

of $V$ are called *letters* or *symbols*, finite strings of letters define words over $V$. The set of the words over $V$ is denoted by $V^*$.

A *generative grammar* $G$ is an ordered fourtuple $(V_N, V_T, S, F)$, where $V_N$ and $V_T$ are finite alphabets with $V_N \in V_T = \emptyset$, $S$ is a distinguished symbol of $V_N$, and $F$ is a finite set of ordered pairs $(P, Q)$ such that $P$ and $Q$ are in $(V_N \cup V_T)^*$ and $P$ contains at least one symbol from $V_N$. The symbols $V_N$ are called *nonterminal* symbols or *variables* and are denoted by capital letters. The symbols of $V_T$ are called *terminal* symbols and are denoted by small letters. The nonterminal symbol $S$ is called the *initial symbol* and is used to start the derivations of the words of the language. The ordered pairs in $F$ are called *rewriting rules* or *productions* and are written in the form $P \rightarrow Q$.

Keeping this formalism, the development of C.e. can be seen as a generative grammar in which the fourtuple $(V_N, V_T, S, F)$ is made respectively of: $S \equiv Zigote$, $V_N = \{S, AB, P_0, P_1, P_2, EMS, \ldots ABa, \ldots ABal, \ldots\}$, with symbols that define cellular stages during development, $V_T = \{a_t, p_t, l_t, r_t, 0\}$ or $V_T = \{ABarppappp, ABalaaaalp, \ldots, 0\}$, if one assume as terminal symbols a special instance of symbols $a, p, l, r$, as used by embryologists, or instead a description through terminally differentiated cells before hatching. In both cases the term 0 for cell death has to be included, and $F = \{AB \rightarrow ABa, ABp; ABa \rightarrow ABar, ABal; \ldots\}$ the rules of production, applied in parallel as in L-systems. Has to be noticed that, from a Biological point of view, the system is not strictly parallel, but for what concern a qualitative description it can be considered as such. The difficult problem to solve is the definition of $F$ if we don't want just a descriptive approach. In fact for decription it is possible to maintain the scheme above with few improvements, but this is not that much informative. Instead if the goal is to be able to describe $F$ in terms of local rules of interactions between genes and cells at the molecular level then the problem is quite complex.

$$
\begin{aligned}
S \equiv (ZIGOTE) &\rightarrow P_0 &;& P_0 &\rightarrow \{AB\}P_1 \\
P_1 &\rightarrow \{EMS\}P_2 &;& \{EMS\} &\rightarrow \{MS\}E \quad (1) \\
P_2 &\rightarrow CP_3 &;& P_3 &\rightarrow DP_4
\end{aligned}
$$

**Founder cells:** $\quad \{AB\}, \{MS\}, E, C, D, P_4 \qquad\qquad (2)$

$$
\begin{aligned}
\{AB\} &\rightarrow \{ABa\}\{ABp\} \\
\{ABa\} &\rightarrow \{ABal\}\{ABar\} \\
\{ABp\} &\rightarrow \{ABpl\}\{ABpr\} \\
\{ABal\} &\rightarrow \{ABala\}\{ABalp\} \\
\{ABar\} &\rightarrow \{ABara\}\{ABarp\} \\
\{ABpl\} &\rightarrow \{ABpla\}\{ABplp\} \\
\{ABpr\} &\rightarrow \{ABpra\}\{ABprp\}
\end{aligned}
\qquad (3)
$$

......

Figure 2: Cell lineages in C. elegans represented as a grammar. The distinguished word $S$ correspond to the Zigote. Every letter, or groups of letters in curly brachets, represent a cell. The first divisions give rise to the founder cells (1,2), subsequent divisions are in general represented by embryologists by addition of an $a$ or $p$ – in case of an anterior, posterior division – or an $l$ or $r$ – for a left, right division –, the beginning of the $\{AB\}$ lineage tree is shown (3). Terminal symbols are in this case the final differentiated cells of the replication tree, or cell death '0' in case of apoptosis.

# 3 Definition of the variable CML

Except when studying some simple cases, as for the alga Anabaena Catenula [5], the modelling proposed by Lindenmayer has not been connected so far with possible underlying biochemical dynamics driving the system, as is the case for gene-expression – see [9] –.

In order to link recent results obtained in simulations of spatially extended chemical systems with CMLs with L-systems, we propose a model for a parametric context-sensitive L-system defined through gene-cell interactions. The basic model has been proposed recently, trough CML formalizations, in order to study the possible interactions in *trans* among genes, through the distribution in time and space of their products among neighbouring cells [1].

The variation in time of the product of a generic gene $G_i$ in a network of $n$ genes, $i = (1, 2, 3, \ldots, n)$, as for example a protein $x_i$ inside a certain cell, can be defined as

$$
x_i(t + 1) = x_i(t)(1 - C_i) + v_i(t)P_i \qquad (4)
$$

where $P_i$ and $C_i$ represent respectively the rate of production and a first order decay for $x_i$. The term $v_i$ is the coupling term for the gene-network of the single cell, which controls the activity of $G_i$, influenced by some other genes, and is defined as a discontinuous function as follow

$$
v_i(t) = \begin{cases} 1, & \text{if } \sum_{j=1}^{n} x_j(t)r_{i,j} \geq k; \\ 0, & \text{otherwise.} \end{cases} \qquad (5)
$$

where $x_j$'s are the concentrations of the gene products present in the system, $r_{i,j}$'s are real numbers representing the interrelationships among them and $k$ represent the threshold value.

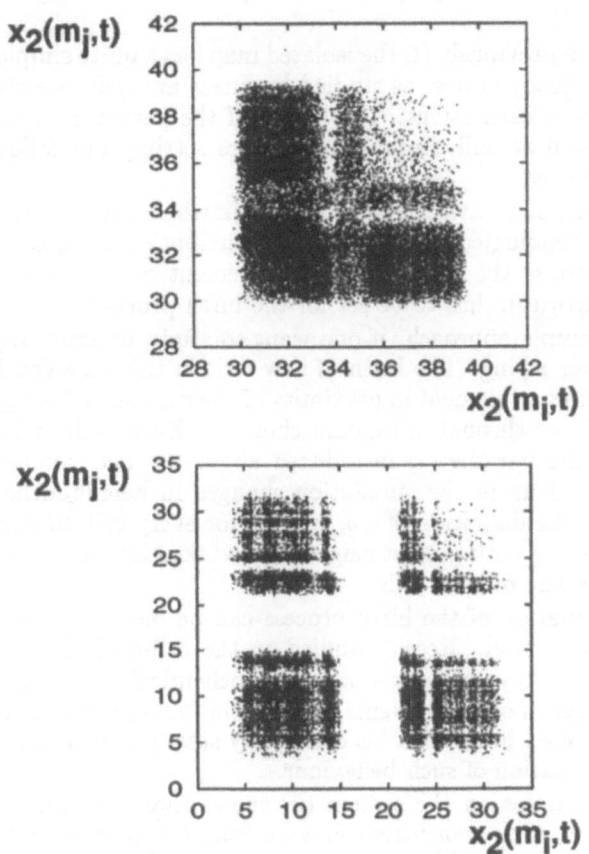

Figure 3: Simulation of a CML dynamics over the ring. Dispersion of the values of the concentrations of the gene – $x_2$ – in a network of two genes in two sites of the lattice – $m_i, m_j$ – over the lattice during a dynamics over the ring. Above: $\mathbf{R} = [-1, 1, 1, -1]$, $C_1 = 0.2$, $C_2 = 0.4$, $P_1 = 3.5$, $P_2 = 19.0$, $\gamma_1 = 0.2$, $\gamma_2 = 0.7$. Below: $\mathbf{R} = [0, 1, 1, -1]$, $C_1 = 0.1$, $C_2 = 0.05$, $P_1 = 3.3$, $P_2 = 5.0$, $\gamma_1 = 0.0$, $\gamma_2 = 0.9$. Lattice size $N = 200$, $T = 3$ for the isolated map in both cases.

Using this simplified description of a cell, the study of the distribution in space of the products, and thus cell-cell interactions, can be done by the addition of a diffusion term in a set of identical equations like the one above, each representing a cell. This procedure gives, for the $i^{th}$ gene of the $m^{th}$ cell of a discrete lattice formed by $N$ identical nearest-neighbour-coupled cells with periodic boundary conditions

$$x_i(m, t+1) = x_i(m, t)(1 - C_i) + v_i(m, t)P_i \ +$$

$$\gamma_i[\sum_{j=1}^{q} x_i(j,t) - qx_i(m,t)] \tag{6}$$

the parameter $\gamma$ above defines the diffusion constant and $q$ the set of the neighbours – e.g. for a ring dynamic with periodic boundary conditions $q = \{m-1, m, m+1\}$ –

As was shown previously [1] the isolated map has a quite simple repertoire of behaviour. All steady states, as studied by linear analysis, are stable, and can be single points or oscillations. The length of the period $T$ of the oscillations can be modulated at will through parameters setting and follows a complex bifurcation sequence.

The definition of a lattice of variable size, in which cell growth occurs, poses immediately a topological problem. The distribution in space of the newly formed cells impose the possibility for movement of neighbouring elements. Moreover an algorithm has to be set for the birth process.

The most simple approach, if one want to study dynamical properties, is the dynamic over a ring. The birth of new cells in this case can be taken into account by adding an element in proximity of the mother cell – e.g. on the right side, or the left, or through a random choice –. Every cell is thus described, in addition to the parameters introduced above, by two pointers setting the neighbourhood. During the simulation changes in neighbourhood are taken into account by the definition of a $q_i$ specific for every cell, moreover the strict structure of the ring with nearest neighbours is taken into account repositioning also pointers for the mother cells.

The determination of the birth process can be based on the characteristic dynamics of the model. Recent studies on the cell-cycle have demonstrated the existence of a very complex system of biochemical interactions inside cells giving rise to a series of cyclic events resulting in cell replication. In this respect the model described here, with his oscillatory steady states can be considered a crude approximation of such behaviour.

Dynamical studies on the system (6) shows that depending on the value of the coupling $\gamma_i$ long transients can arise from the coupled dynamics among elements over the lattice. A typical behaviour for the case of a period $T = 3$ attractor for the isolated network and for a long transient over the lattice dynamic is shown in Fig. 3. One can thus establish, similarly to what is thought to happen in-vivo, that a certain range of concentrations, is able to switch replication. A further set of parameters: $l_i(min), l_i(max)$, set the range limits that trigger such behaviour, for any number of gene products in the network.

Results obtained with rings dynamics starting from random initial conditions for a ring of $N = 20$ elements, and allowing the ring to growth up to $N = 1000$ elements are shown in Fig. 4 and 5.

Depending from the choice of the swithching threshold, from the parameters for the isolated map, and from the coupling, different dynamics arise, with different speeds of growth. This approach allows for the study of the dynamics in the presence of variations in space induced by changes in the degrees of freedom of the system.

In the case of two or three dimentional dynamics the topological problem is obviously more difficult to solve. Lattice growth in this case has to take into

Figure 4: Growth dynamics over the ring, see text for explanations.

account the presence of a more complicated set of interactions surrounding every cell.

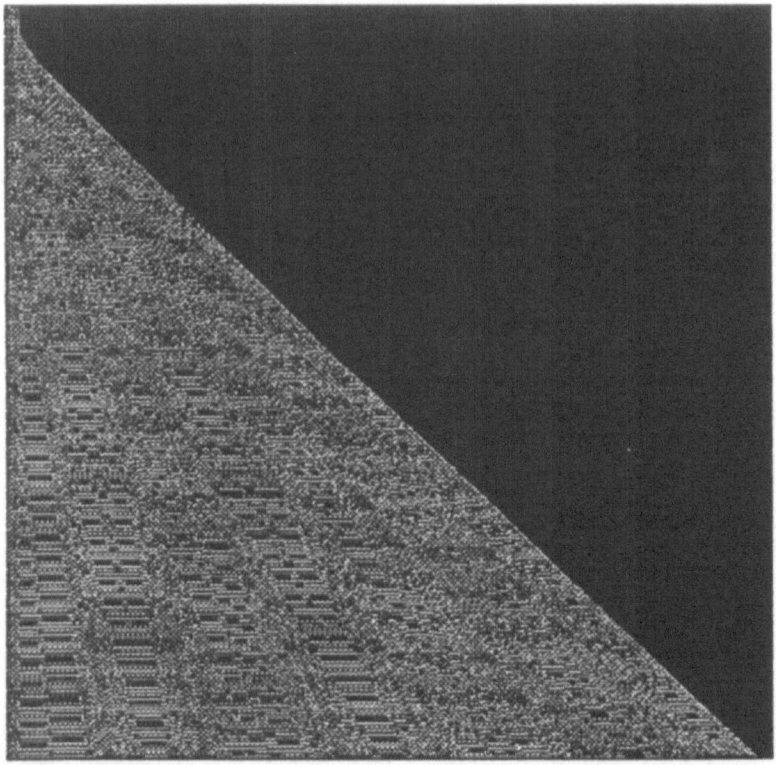

Figure 5: Growth dynamics over the ring, see text for explanations.

The possible solutions in this case are quite different depending on the use of the model. In the case of studies related to the analysis of the dynamics of the system a mixed approach similar to that implemented with Lattice Gas Cellular Automata [LGCA] whith CML's that move reciprocally can be used. In this case cells sit on a grid where they can move, and on which they can replicate. Further possibilities of simulation can be implemented using CAs or PDEs, as was reported in the introduction.

# 4 CMLs as L-systems

In this context we have to refer to parametric OL-systems. These operate on *parametric words*. In this case the alphabet $V$ is made of *modules* wich include a letter , e.g. $X \in V$, and parameters $x_1, x_2, x_3, \ldots, x_n \in \mathbf{R}$ associated to $X$.

A parametric OL-system is thus defined as an ordered quadruplet $G = \langle V, \Omega, S, F \rangle$, where: $V$ is the *alphabet* of the system, $\Omega$ is the set of formal parameters, $S$ is a nonempty word, $F$ is a finite set of productions.

A further complication necessary to keep with the biological description imply the introduction of local interactions, the system become thus a parametric OL-system context sensitive in such case the productions have to be of the type

$$X_{m-1}(x_1,\ldots,t)X_m(x_1,\ldots,t)X_{m+1}(x_1,\ldots,t) \rightarrow$$
$$X_{m-1}(x_1,\ldots,t+1)X_m(x_1,\ldots,t+1)X_{m+1}(x_1,\ldots,t+1) \qquad (7)$$

in which the state of the $X_m(\ldots t+1)$ element is determined by its parameters at time $(t)$ and by its neighbours.

We can now rephrase our CMLs in the following way: we define a generative grammar context sensitive, with parallel updating, in which the alphabet $V$ is defined by $V = \{00, 11, 01, 10\ldots\}$ – i.e. we define the alphabet as the possible states of our genetic networks corresponding to $v_i(t)$ in equation (5), the numerosity of our alphabet will be defined by $2^n$ where $n$ correspond to the number of genes present in the isolated cell, $S$ correspond to an initial set of genes i.e. $S = \{001, 111, 001, 110, 001\}$ for a group of five cells with three genes each, while $\Omega = \{x_i, r_{ij}, v_i, k_i, l_i(max), l_i(min), C_i, P_i, \gamma_i\}$ is the set of parameters, as defined in equation (6), and $F$ the productions. The productions will be set following eq. (6) but including rules for replication of the type

$$X_{m-1}(001, x_i\ldots)X_m(110, x_i\ldots)X_{m+1}(110, x_i\ldots) \rightarrow$$
$$X_{m-1}(001, x_i\ldots)X_m(110, x_i\ldots)X_{daughter}(110, x_i\ldots)X_{m+2}(110, x_i\ldots)(8)$$

We can define in this way functional states of cells, or cells in cell lineages in terms of gene activation. The organism is thus defined in its components by the differentiative states, or embryonal states, of its cells. Once a set of arbitrary defining words $V^*$ is set the resulting language $L$, which include cell states, will define the organism, and some of its dynamics, in a formal manner.

# 5  Results and discussion

The modelling presented above can be used for the study of the dynamical behaviour of systems which shows a variation in the number of degrees of freedom. The study of such models is still an open field. This imply that a quite large body of preliminary work is necessary in order to study these problems from the perspective of dynamical systems theory. In this respect the model, grownded on an experimental basis derived from the study of biological systems, is simple enought to allows a detailed theoretical study of its long term dynamical behaviour.

From the point of view of the study of Biological systems, the model presented here is particularly suited for the study of C.e. development, but it adopt a very general point of view. Because of this fact it can be used as a tool to validate working hypothesis in a broad range of different situations, or as a conceptual frame for rationalization in the case of data collections.

In the case of C.e. its usefulness is based on the fact that: cell lineages – i.e. the detailed pattern of cell replications –, cell-cell interactions – expression of specific genes that set embryonic development –, and cellular positions – through time-lapse cinematography studies –, are all available in this particular model. This mean that it is possible to trace all the lineages with: cell divisions, cell number, and cell movement. This give rise to a complex structure of a dynamic binary tree, in three dimentions, representing development.

Theoretical modelling of this behaviour and testing known hypothesis based on genetic findings, is now possible because of this particularly useful setting. Moreover, results obtained throught simulations, thanks to the possibility of genetic manipulations, can be tested through experiments also in their prediction capability. The full picture of C.e. development is still being developed, but has now reached a critical mass already amenable for modelling.

At the same time, because of the huge complications present in Biological systems, a word of caution is obviously necessary. The implementation of particulars, in terms of molecular interactions, will obviously require sets of different models, in order to understand fully the dynamics of the possible interactions inside the system. This will pose most likely new questions, one of which is to which extent will we be able to pursue such analysis.

For example all variables described in

$$\Omega = \{x_i, r_{ij}, v_i, k_i, l_i(max), l_i(min), C_i, P_i, \gamma_i\} \tag{9}$$

are complex variables in vivo, the same holds for topological variables, not discussed extensively here. This imply a quite large body of work for a correct definition. At the same time, an ensamble view is important. The main reason being the possibility to get a set of general considerations drawn from the type of modelling described here. The approach that one has to follow to obtain this result is to study separately the problems present in the different subsets knowing complications that can arise out of the global picture.

As was already mentioned this work has also a quite general value, the extension to other systems is obviously possible following a similar rationale. This fact can have a quite large impact also in other fields of research in Biology, in which the problems are similar. What is a drawback in perpsective is the fact that other experimental models do not allow at the moment such a detailed description because of the intimate nature of their dynamics, or because of their size.

In order to clarify this point of view an example can be given, based on simulations done so far. If one consider dynamics over the ring for the CML described in (6) the isolated map is stable by construction, this mean that the only possible states can be stable points or oscillations. Despite this 'simple' behaviour the coupling can give rise to long term dynamics with transients that increase with lattice size, the system is thus bound to settle into two different main modes of behaviour in terms of long terms dynamics depending on the coupling $\gamma_i$.

For a certain subset of coupling strenghts the final steady state is reached in a relatively short time, while for a certain region of parameter space transients growth exponentially with lattice size. For systems of sufficient sizes, as is the case for several biological systems, there is thus the possibility of locking into a never ending transient dynamics away from the final steady state.

Another interesting aspect in the definition of $\gamma_i$, and in this aspect of transient dynamics is the fact that in-vivo the transmission of information, inside cells and among them, is done through a complex set of biochemical interactions generally indicated as *"signal transduction pathways"*. The high complex molecular interrelations present in these systems could drive the dynamics in different regions of parameter space modulating in this way the transients. Local signal transmission can thus by itself give rise to a quite complex pattern of behaviours, aside from other possible mechanisms.

# References

[1] BIGNONE, F.A., Cells-Gene interactions simulation on a coupled map lattice, *Journal of Theoretical Biology*, **161**(2), 231-249, 1993.

[2] COSENZA, M.G., KAPRAL, R., Coupled maps on fractal lattices, *Physica Review A*, **46** (4), 1850-1858, 1992.

[3] KAPRAL, R., Discrete models for chemically reacting systems, *J. Math. Chem.*, **6**, 113-163, 1991.

[4] HASTINGS, A., HIGGINS, K., Persistence of transients in spatially structured Ecological Models, *Science*, , 1133-1136, 1994.

[5] KOSTER, C.G., LINDENMAYER, A., Discrete and continuous models for heterocyst differentiation in growing filaments of blue-green bacteria, *Acta Biotheoret.*, **36**, 249-273, Kluwer Academic, the Netherlands, 1987.

[6] LINDENMAYER, A., Mathematical models for cellular interaction in development, Parts I and II, *J. Theoret. Biol.*, **18**, 280-315, 1968.

[7] PREHN, R.T., Many growth factors may not be growth factors, *Cancer Res.*, **52**, 501-507, 1992.

[8] PRUSINKIEWICZ, P., HANAN, J., Visualization of botanical structures and processes using parametric L-systems. In Thalmann, D., editor, *Scientific Visualization and Graphic Simulation*, 183-201, J. Wiley & Sons, 1990.

[9] PRUSINKIEWICZ, P., LINDENMAYER,. A., *The algorithmic beauty of plants*, Heidelberg & New York, 1990.

[10] RAFF, M.C., BARRES, B.A., BURNE, J.F., COLES, H.S., ISHIZAKI, Y., JACOBSON, M.D., Programmed cell death and the control of Cell survival: lessons from the nervous system, *Science*, **262**, 695-700, 1993.

[11] ROZENBERG, G. SALOMAA ,. A., The mathematical Theory of L systems, New York & London, 1980.

[12] WOOD, W.B., The Nematode Caenorhabditis elegans, Cold Spring Harbour Labs., N.Y., 1988.

**Acknowledgements:** the author wish to thank Roberto Livi, Antonio Politi, Salvatore Di Gregorio, Ralf and Heinke Schnabel for helpful discussions. Thanks to Ralf Schnabel for Fig. 1. Simulations were done on a Personal IRIS, Silicon Graphycs Inc., HP740, Hewlett Packard Inc., and IBM-RISC6000, IBM Inc. Fig. 4 and 5 have been obtained through data formatting using the *ximage HDF* – Hierarchical Data Format – package, from the National Center for Supercomputing Applications at the University of Illinois in Urbana Champaign, Illinois, U.S.A. This work has been partially supported by C.N.R. grant # 95.01751.CT14 "Studio analitico della dinamica della regolazione genica e della morfogenesi".

# Essential Transformations of the One Dimensional Cellular Automata Rule Space and Endomorphisms of Compact Abelian Groups *

G. Cattaneo          E. Formenti          G. Mauri
A. Vaccaro
Dipartimento di Scienze dell'Informazione
39 Via Comelico, 20135 Milano, Italy.

L. Margara
Dipartimento di Scienze dell'Informazione
7 Via Mura Anteo Zamboni, Bologna, Italy.

### Abstract

In this paper we review some of the most popular definitions of chaos. In order to distinguish simple shift-like dynamics from one-sided shift-like dynamics we introduce CF-chaotic systems. We prove that this class is not empty and we study some correlations with the other chaotic behaviors. Moreover in the second part of the paper we prove that essential transformations of the cellular automata rule space preserve the global qualitative dynamics when considering endomorphisms of compact abelian groups. We stress that the chaotic behaviors (with respect to the definitions given in the former part of paper) are preserved too.

## 1  Introduction

Cellular automata (CA from now on) are computing devices that have been used in many research fields, mainly for simulating natural phenomena. The evident chaoticity of most of natural phenomena greatly developed the study of the "chaotic" behavior of some CA. Moreover the study of the CA behavior became a central topic of discrete dynamical systems theory. Since there is no universal accepted definition of chaotic system, in this paper we briefly review the most used ones (Devaney,Knudsen) and we propose a new definition. The main purpose of our definition is to distinguish the class of chaotic systems that are topologically conjugate to biinfinite shift from those that are topologically conjugate to one sided shifts. We termed systems in the latter class as *CF-chaotic*. We prove that this class is not empty since at least LRCA are CF-chaotic. Moreover we give a classification of chaotic elementary CA.

*This work has been partially supported by MURST, under the project "Efficienza di Algoritmi e Progetto di Strutture Informative", by CEC ESPRIT BRA contract n. 6317 - ASMICS 2 and by CNR contract n. CT94.00452.CT12.115.26124

In general, proving that a CA has a certain dynamical behavior may be quite difficult. For this reason in [3] we have introduced some groups of transformations of the rule space and we prove that they preserve some general dynamical properties. In this way the rule space is partitioned in equivalence classes. All the CA in the same equivalence class have the same qualitative behavior. Then if we compile a dynamical classification we have to study the behavior of only one representative per class.

In this paper we review the class of essential transformations and we prove new results for the special case of CA that are endomorphisms of compact abelian groups. In this case we prove that essential transformation preserve all the most interesting dynamical properties such as: ergodicity, transitivity, expansivity, sensitivity to initial conditions, denseness of periodic points. The proofs we give hold only for the topological group $Z_2$, but in the last section we generalize some of these results.

The paper is structured as follows. Section 2 gives some general definitions about CA. In section 3 we briefly review the main definitions of chaotic systems given in literature and we introduce CF-chaotic systems. Moreover we prove that strongly transitive CA are surjective but not injective and that the class of CF-chaotic CA is not empty since it contains at least $\mathcal{LRCA}$. Then we classify $\mathcal{ECA}$ from the point of view of chaotic behavior. We also prove that $\mu(\Pi) = 0$ does not characterize a chaotic CA. In section 4 and 5 we review some known results on transformations of the rule space. In section 5.1 we fully characterize the properties preserved by essential transformations when the rule space is restricted to endomorphisms of compact abelian groups over $Z_2$ and their essential transformed rules. In section 6 some results of section 5.1 are generalized to groups of prime cardinality.

# 2 Cellular automata

A one dimensional CA is a triple $\langle S, N, f \rangle$, where $S = \{0, 1, \ldots, p-1\}$ is the set of *states*, $N = \{-r, \ldots, 0, \ldots, r\}$ is the *neighborhood* structure with *rule radius* $r$, and $f: S^{2r+1} \to S$ is the *local rule*. For any fixed set of states $S$ and any rule radius $r$ the *rule space* is the set, denoted by $\mathcal{R}(S, r)$, of all the local rules $f: S^{2r+1} \to S$. In the sequel, for the sake of simplicity, we will often use elementary CA. A one dimensional CA is elementary if $S = \{0, 1\}$ and $r = 1$. Let $\mathcal{ECA}$ be the set of all elementary CA. Each elementary rule $f: \{0, 1\}^3 \to \{0, 1\}$ is bijectively associated with the $\{0, 1\}$-valued vector of $2^3$ components:

$$\xi_f \equiv (f(1, 1, 1), f(1, 1, 0), \ldots, f(0, 0, 0)) \in \{0, 1\}^{2^3}$$

called the $\{0, 1\}$-*vector representation* of $f$. Each elementary rule $f$ can be encoded with the integer number $R_f$ whose base 2 representation is $\xi_f$. This procedure can be easily extended to non elementary one dimensional CA, representing any rule $f \in \mathcal{R}(S, r)$ by an $S$-valued vector (table) $\xi_f$ of $p^{2r+1}$ components. The *global function* of a given CA with local rule $f$ is the mapping $G_f : S^{\mathbf{Z}} \mapsto S^{\mathbf{Z}}$ associating to any bi-infinite $S$-valued sequence $c : \mathbf{Z} \mapsto S$, $i \to c(i)$ the bi-infinite $S$-valued sequence $G_f(c) : \mathbf{Z} \mapsto S$, $i \to G_f(c)(i)$ defined as follows:

$$G_f(c)(i) = f(c(i - r), \ldots, c(i), \ldots, c(i + r)).$$

As an example, the global function of the *shift* CA, denoted by $\sigma$, with radius $r \geq 1$ and $S$ as its set of states is defined $\forall c \in S^{\mathbf{Z}}, \forall i \in \mathbf{Z}$ by $\sigma(c)(i) = c(i+1)$.

The collection $S^{\mathbf{Z}}$ of all bi-infinite $S$-valued sequences is the *phase space* of the discrete time dynamics generated by iterations of the global function $G_f$; the elements of $S^{\mathbf{Z}}$ are said to be *configurations*. The mapping $d: S^{\mathbf{Z}} \times S^{\mathbf{Z}} \to \mathbb{R}^+$ defined by:

$$\forall x, y \in S^{\mathbf{Z}} \quad d(x, y) = \sum_{i=-\infty}^{+\infty} \frac{\delta(x_i, y_i)}{2^{|i|}}$$

(where $\forall a, b \in S$, $\delta(a, b) = 1$ if $a \neq b$, 0 otherwise) is a metric on the phase space $S^{\mathbf{Z}}$, whose corresponding topology coincides with the product topology induced by the discrete topology of $S$. With this topology $S^{\mathbf{Z}}$ is a Cantor space, i.e. a compact, totally disconnected and perfect space. A subbase of clopen (i.e. closed and open) sets for $S^{\mathbf{Z}}$ consists of all the sets of the form $S_i(s) = \{c \in S^{\mathbf{Z}} \mid c(i) = s\}$ with $s \in S$. Then every open set is a finite intersection of elements of the subbase. Let us recall that with respect to this topology any global function $G_f$ generated by a local rule $f$ is a continuous mapping.

Hedlund ([9],Thm. 3.4) proved that the following two statements are equivalent.

1. $F : S^{\mathbf{Z}} \mapsto S^{\mathbf{Z}}$ is continuous and commutes with the shift map (i.e. $F \circ \sigma = \sigma \circ F$).

2. $F : S^{\mathbf{Z}} \mapsto S^{\mathbf{Z}}$ is a global function of a suitable local rule $f$, i.e., $F = G_f$.

A CA is injective (surjective or open, respectively) if and only if its global function is injective (surjective or open, respectively).

# 3 Chaos in cellular automata

The concept of chaos is very appealing for the scientists. Unfortunately, in literature, there in no universal accepted definition of deterministic chaos. One of the most used definitions is due to Devaney [6].

**Definition 1** *Devaney's chaotic DTDS*
*A DTDS is Devaney chaotic if it is sensible to initial conditions, regular, topologically transitive.*

In [1] Banks et al. proved that the above conditions are not independent; in fact 2) and 3) imply 1). The main drawback of this definition is that if you have a Devaney chaotic system then its restriction to periodic orbits is Devaney chaotic too. This fact induced Knudsen to modify the definition of chaotic system as follows.

**Definition 2** *Knudsen's chaotic DTDS*
*A DTDS is Knudsen chaotic if it is sensible to initial conditions and has a dense orbit.*

Recall that, in the case of compact phase spaces, a DTDS is transitive if and only if it has a dense orbit. Moreover in [5] it is proved that, as for CA, transitivity implies sensibility to initial conditions. Then the definition of Knudsen chaotic CA reduces to the following.

**Proposition 1** *A CA is Knudsen chaotic if and only if has a dense orbit.*

The shift map is a paradigmatic example of both Devaney and Knudsen chaotic system. A very interesting class of Devaney chaotic systems is that of leftmost and rightmost permutive CA (briefly $\mathcal{LRCA}$).

**Definition 3** *A CA is permutive in the $i_{th}$ variable $(-k \leq i \leq k)$ if and only if for any sequence $x_{-r}, \ldots, x_{i-1}, x_{i+1}, \ldots, x_r \in S^{2r}$ we have*

$$\{f(x_{-r}, \ldots, x_{i-1}, x_i, x_{i+1}, \ldots, x_r) : x_i \in S\} = S.$$

**Definition 4** *Leftmost permutive CA*
*Let $f$ be a CA local rule. The CA is leftmost (rightmost) permutive if there exists an integer $i$, $-r \leq i \leq r$ such that*

- *$i < 0$ $(i > 0)$*

- *$f$ is permutive in the $i^{th}$ variable*

- *$f$ does not depend on $x_j, j < i$ $(j > i)$.*

A CA is in $\mathcal{LCA}$ $(\mathcal{RCA})$ if it is leftmost (rightmost) permutive. Then $\mathcal{LRCA} = \mathcal{LCA} \cap \mathcal{RCA}$. Moreover let $\mathcal{ICA} = \left\{ G_f \mid \exists n \in \mathbb{N}, \ G_f^n = Id \right\}$. The following theorem gives a relation between $\mathcal{LRCA}$ and chaos.

**Theorem 1 [7]** *$\mathcal{LRCA}$ are both Knudsen and Devaney chaotic.*

If we consider the space-time diagram of a $\mathcal{LRCA}$ and the one of a shift-like CA, we notice that they are completely different. Moreover the CA in the former class seem to fit much better the intuitive concept of chaos even from a "pictorial" point of view. A property that seems to distinguish the two classes is expansivity. But in [8] it is proved the following.

**Theorem 2** *There are no weakly expansive CA in dimension higher than one.*

As an easy consequence we have that there are no expansive CA in dimension higher than one. This means that if we use expansivity in defining chaotic phenomena in CA then the definition is empty for $d > 1$. We have to search for less trivial properties. Strongly transitivity is another interesting property that is not shared between full shift-like behavior and one sided shift-like behavior. Then we can give the following.

**Definition 5** *A DTDS is CF chaotic if it is completely mixing and strongly transitive.*

We remark that definition 5 is a topological definition since both completely mixing and strongly transitivity are topological properties. In the case of CA in [2] it is proved the following:

**Theorem 3** *Topologically transitive CA with respect to the metric topology induced by the Tychonoff distance are ergodic with respect to the normalized Haar measure.*

Since a CA is completely mixing if and only if it is ergodic and since strongly transitivity implies topological transitivity, by theorem 3, a CA is CF-chaotic if it is strongly transitive.

We remark that the full shift over an alphabet $S$ can not be strongly transitive. The property that makes the shift not strongly transitive is injectivity, as it is proved in the following theorem.

**Theorem 4** *Strongly transitive CA are surjective but not injective.*

**Proof** — Let $G_f$ be a strongly transitive CA but not surjective. Then we have:

$$\bigcup_{n \in \mathbb{N}} G_f^n(S^{\mathbb{Z}}) \subseteq G_f(S^{\mathbb{Z}}) \subset S^{\mathbb{Z}}.$$

This is an absurd since, by strongly transitivity, we should have

$$\bigcup_{n \in \mathbb{N}} G_f^n(S^{\mathbb{Z}}) = S^{\mathbb{Z}}.$$

Now let us prove the second part of the theorem. Fix an $\epsilon > 0$. Pick $x, y \in S^{\mathbb{Z}}$ then, because of strongly transitivity, there exists $z \in U_\epsilon(x)$ and $m \in \mathbb{N}$ such that $G_f^m(z) = y$, where $U_\epsilon(x)$ is an open neighborhood of $x$ of diameter $2\epsilon$. Take $x_0 \in S^{\mathbb{Z}}$ such that $d(x, x_0) = 2\epsilon$. By the strongly transitivity of $G_f$, we have that there exists $z_0 \in U_{\frac{\epsilon}{2}}(x_0)$ and $n \in \mathbb{N}$ such that $G_f^n(z_0) = y$. Since $0 \leq \frac{1}{2}\epsilon \leq d(z, z_0) \leq \frac{7}{2}\epsilon$, we have that $z \neq z_0$. ●

Theorem 4 explains the reason we have chosen strongly transitivity to distinguish the full shift-like chaos from the harder form of chaos that is typical of systems topologically conjugated to one sided shifts.

From theorem 4 we trivially deduce that CF-chaotic CA are surjective but not injective, it is an open problem to see if all CA with these properties are CF-chaotic.

Now we are going to prove that the class of CF-chaotic CA is not empty. To this purpose we use a result in [7]:

**Theorem 5** ([7]) $\mathcal{LRCA}$ *are topologically conjugated to a one sided shift over a suitable alphabet $S$.*

The following theorem is a slight modification of a result in [11].

**Theorem 6** *Let $\sigma_S$ be a one sided shift over the alphabet $S$. Then $\sigma_S$ is strongly transitive.*

**Proof** — Let $x \in S^{\mathbb{Z}}$ and let $U_\epsilon$ be an open neighborhood of $x$ of diameter $2\epsilon$. Then $\exists n \in \mathbb{N} : \forall y \in U_\epsilon$, $x_i = y_i$ with $0 \leq i \leq n$. Surely there exists $z \in U_\epsilon$ such that

$$\forall i \in \mathbb{N}, \; z_i = \begin{cases} x_i & \text{if } 0 \leq i \leq n, \\ y_{i-n} & \text{otherwise.} \end{cases}$$

Then $\sigma_S^n(z) = y$. ●

From theorem 6 and 5 we have the following.

**Corollary 1** $\mathcal{LRCA}$ *are strongly transitive.*

From corollary 1 and theorem 5 we have the following.

**Proposition 2** $\mathcal{LRCA}$ *are CF-chaotic.*

Proposition 2 proves that the class of CF-chaotic CA is not empty. It is an open problem to see if all CF-chaotic CA are Devaney chaotic or not.

**Theorem 7 [12]** $\mathcal{LCA} \cup \mathcal{RCA}$ *are strongly mixing and hence ergodic.*

As for $\mathcal{ECA}$, we have that $\mathcal{SCA} = \mathcal{LCA} \cup \mathcal{RCA} \cup \mathcal{ICA}$. From theorem 7 we may conclude that an elementary CA is ergodic if and only if it belongs to $\mathcal{SCA} \setminus \mathcal{ICA}$. We have proved the following.

**Theorem 8** *As for* $\mathcal{ECA}$, *all CA in* $\mathcal{SCA} \setminus \mathcal{ICA}$ *are strongly mixing and hence ergodic.*

It is worthwhile of remarking that surjectivity is a decidable property.

Let $\Pi$ be the set of periodic points of a DTDS. Sometimes the fact that $\mu(\Pi) = 0$ is considered significative feature of the presence of chaos. This idea is misleading as for CA. In fact we have

**Theorem 9** *All CA in* $\mathcal{ECA} \setminus \mathcal{ICA}$ *are such that* $\mu(\Pi) = 0$.

**Proof** — If $G_f$ is not surjective then $\mu(\Omega_f) = 0$ and this implies that also $\mu(\Pi) = 0$. Theorem 8 says that $\mathcal{SCA} \setminus \mathcal{ICA}$ are strongly mixing, but this implies $\mu(\Pi) = 0$. $\bullet$

Even if $\mu(\Pi) = 0$, CA may nonetheless have many periodic points; in fact, from a corollary of theorem 3 in [11], a CA in $\mathcal{ACA} \cap (\mathcal{LCA} \cup \mathcal{RCA})$ has a dense set of periodic points (for the definition of $\mathcal{ACA}$ see section 5.1).

# 4 Transformations of CA rule spaces

Given a set $\mathcal{X}$, a *transformation* of $\mathcal{X}$ is any mapping $F: \mathcal{X} \mapsto \mathcal{X}$ which is both injective and surjective. Transformations defined over finite sets are called *permutations*. Let $\mathcal{T}(\mathcal{X})$ be the set of all the permutations of $\mathcal{X}$ and $\circ$ the usual operation of function composition. One can easily verify that the pair $\langle \mathcal{T}(\mathcal{X}), \circ \rangle$ is a group (the *symmetric group* of $\mathcal{X}$) and that every subset of $\mathcal{T}(\mathcal{X})$ closed with respect to $\circ$ is a subgroup of $\langle \mathcal{T}(\mathcal{X}), \circ \rangle$.

Given a permutation $\tau \in \mathcal{T}(\mathcal{X})$, an element $x \in \mathcal{X}$ is *moved* by $\tau$ if $\tau(x) \neq x$, otherwise the element is *fixed* or is a *fixed point* of $\tau$. Two permutations $\tau_1$ and $\tau_2$ are *disjoint* if and only if the intersection between the set of elements moved by $\tau_1$ and the set of elements moved by $\tau_2$ is empty. Two disjoint transformations commute with respect to $\circ$. Moreover if the elements of a subgroup $\langle \mathcal{G}, \circ \rangle$ of $\langle \mathcal{T}(\mathcal{X}), \circ \rangle$ are pairwise disjoint then $\langle \mathcal{G}, \circ \rangle$ is abelian.

Let $\mathcal{G} \subseteq \mathcal{T}(\mathcal{X})$ be a subgroup, we define $\mathcal{R}_{\mathcal{G}}$, the *relation induced by* $\mathcal{G}$ on $\mathcal{X}$, as follows:

$$\forall x, y \in \mathcal{X} \quad x \mathcal{R}_{\mathcal{G}} y \Leftrightarrow \exists \tau \in \mathcal{G} \text{ such that } \tau(x) = y.$$

It is easy to see that $\mathcal{R}_{\mathcal{G}}$ is an equivalence relation on $\mathcal{X}$. Let $F(\tau)$ be the set of fixed points of the transformation $\tau \in \mathcal{G}$, that is $F(\tau) = \{x \in X \mid \tau(x) = x\}$.

The concept of transformation (i.e. permutation) can be specialized to the case of arbitrary (any radius and any finite set of states) CA rule spaces.

**Definition 6** *Transformation of CA local rules*
*Let $S$ be a finite set of states and $r$ a local radius, then a transformation of the space of $(S, r)$-CA local rules is a one-to-one transformation from $\mathcal{R}(S, r)$ onto $\mathcal{R}(S, r)$.*

The set of all transformations of the CA rule space $\mathcal{R}(S, r)$ will be denoted by $\mathcal{T}(\mathcal{R}(S, r))$. Let $h$ be a self-inverse permutation of $S$ and $\underline{h}: \mathcal{R}(S, r) \to \mathcal{R}(S, r)$ such that $\forall f \in \mathcal{R}(S, r)$, $\underline{h}(f) = (h \circ f): S^{2r+1} \to S$. We say that $\underline{h}$ is the transformation of the rule space induced by the self-inverse permutation $h$. Let $\mathcal{H}$ be the set of all the transformations of the rule space induced by self-permutations of $S$.

In particular we distinguish the identical transformation $h_i : \mathcal{R}(S, r) \to \mathcal{R}(S, r)$ where $h_i(f) = f$, and the complement transformation $h_c : \mathcal{R}(S, r) \to \mathcal{R}(S, r)$, where $h_c(f): S^{2r+1} \to S$ is defined $\forall x \in S^{2r+1}$ by the law $h_c(f)(x) = \overline{f(x)}$ where $\forall s \in S$, $\bar{s} = p - 1 - s$.
On the other hand, let $\mathcal{G}$ be the set of all permutations of $S^{2r+1}$. For any permutation $g$ of $S^{2r+1}$, we may construct a transformation $\tau_g : \mathcal{R}(S, r) \to \mathcal{R}(S, r)$ of the rule space $\mathcal{R}(S, r)$ associating to the CA rule $f : S^{2r+1} \to S$ the CA rule $\tau_g(f) = (f \circ g) : S^{2r+1} \to S$.

**Definition 7** *Double permutation of CA*
*A transformation of CA rule space $\mathcal{R}(S, r)$ is a double permutation if and only if it is the composition of two permutations $g \in \mathcal{G}$ and $h \in \mathcal{H}$, associating to the rule $f$ the new rule $\tau(g, h)(f) := h(f \circ g)$.*

A transformation $\tau \in \mathcal{T}(\mathcal{R}(S, r))$ *preserves* a property $\mathcal{P}$ if and only if every CA $f$ that satisfies $\mathcal{P}$ is such that $\tau(f)$ satisfies $\mathcal{P}$ too.

# 5   Essential transformations

*Essential transformations* have been introduced by Sutner [13]. In this section we briefly review some known results [3] that will be useful later. In [3] it is proved that essential transformations preserve the complexity of the language associated with the CA. Moreover they preserve set properties such as surjectivity, injectivity and openness.

Consider the following transformations of CA, for all $f \in \mathcal{R}(S, r)$, let $\tau_i, \tau_e, \tau_c, \tau_{ce}$ be defined as follows:

$$\forall (x_{-r}, \ldots, x_0, \ldots, x_r) \in S^{2r+1}$$

$$
\begin{aligned}
\tau_i(f)(x_{-r}, \ldots, x_0, \ldots, x_r) &= f(x_{-r}, \ldots, x_0, \ldots, x_r) \\
\tau_c(f)(x_{-r}, \ldots, x_0, \ldots, x_r) &= \overline{f(x_{-r}, \ldots, x_0, \ldots, x_r)} \\
\tau_e(f)(x_{-r}, \ldots, x_0, \ldots, x_r) &= f(\overline{x_{-r}}, \ldots, \overline{x_0}, \ldots, \overline{x_r}) \\
\tau_{ce}(f)(x_{-r}, \ldots, x_0, \ldots, x_r) &= \overline{f(\overline{x_{-r}}, \ldots, \overline{x_0}, \ldots, \overline{x_r})}.
\end{aligned}
$$

All the above transformations are double permutations. It is easy to show that $\tau_e$ and $\tau_c$ are pairwise disjoint and that the class $\mathcal{T}_e = \{\tau_i, \tau_e, \tau_c, \tau_{ce}\}$ is closed with respect to $\circ$. Then the induced relation $\mathcal{R}_e$ is an equivalence relation.

**Proposition 3 [3]** *Let* $k = p^{2r+1}$ *then:*

$$\left| \mathcal{T}(\mathcal{R}(\mathcal{S}, r))/_{\mathcal{R}_e} \right| = \begin{cases} \frac{p^k + 2 \cdot p^{\frac{k}{2}}}{4} & \text{if } p \text{ is even} \\ \frac{p^k + 2 \cdot p^{\frac{k+1}{2}}}{4} & \text{otherwise.} \end{cases}$$

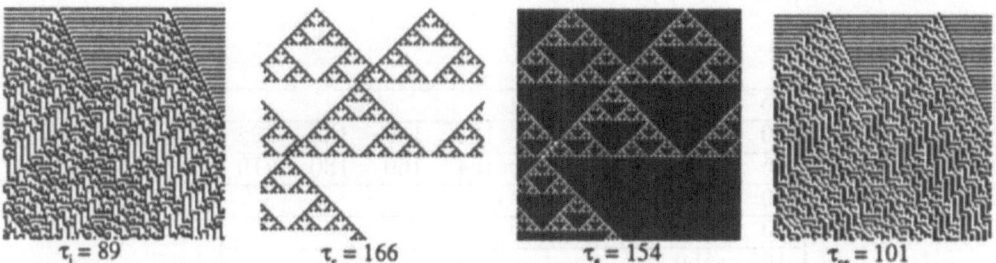

$$\tau_i = 89 \qquad \tau_c = 166 \qquad \tau_e = 154 \qquad \tau_{ce} = 101$$

Figure 1: Example of space-time patterns for the class of rule 89, starting from a configuration $x$ (89, 154) and from $\bar{x}$ (101, 166).

In what follows we prove theorems only for $\tau_c$. In fact, in [3] $\tau_{ce}$ is proved to be an isomorphism. Then if a topological or metrical property is preserved by $\tau_c$ and $\tau_{ce}$, it is preserved by $\tau_c \circ \tau_{ce} = \tau_e$ too.

As stressed in [3] there is no "*morphism*" of dynamical systems from $\tau_i(f)$ to $\tau_e(f)$, for a given $f \in \mathcal{R}(\mathcal{S}, r)$. This is well underlined by the following example.

**Example 1 [3]** Consider the elementary rule 128. The transformed rule under $\tau_c$ has code 127. Suppose that there exist a $H: \{0, 1\}^{\mathbb{Z}} \to \{0, 1\}^{\mathbb{Z}}$ such that $H \circ G_{128} = G_{127} \circ H$. Let us consider the quiescent configuration $\underline{0}$, then:

$$H(\underline{0}) = G_{127}(H(\underline{0}))$$

that is $H(\underline{0})$ is a fixed point of $G_{127}$. Absurd, since $G_{127}$ has no fixed points! (The claim can be proved by using the methods explained in [4]). Then the rules 128 and 127 are not topologically conjugate.

Figure 1 underlines the results in example 1. Here the space-time patterns of rule 89 and 154 are completely different. There are some special cases where essential transformation preserve the global qualitative dynamics. This is the case of additive rules as we are going to see in the next section.

Essential transformations are important, also, from a measure theoretic point of view. In order to explain this fact we have to state the following.

**Theorem 10 [12]** *A CA is surjective if and only if it is measure preserving.*

Since the class of essential transformations preserve surjectivity, they also preserve the property of being measure preserving.

**Proposition 4** *Let $\tau \in \mathcal{T}_e$. If $G_f$ is a global function of a one dimensional CA that preserves the measure then $G_{\tau(f)}$ preserves the measure too.*

**Remark 1** *As a consequence of proposition 8 we have that the class of minimal transformations ([3]) preserve ergodicity.*

Moreover it is very easy to prove the following.

**Proposition 5** *The class of essential transformations preserve the property of being leftmost (rightmost) permutive.*

| 15 | 85 | 170 | 240 | | | | |
|----|----|----|----|----|----|----|----|
| 30 | 86 | 106 | 120 | 135 | 149 | 169 | 225 |
| 45 | 75 | 89 | 101 | 154 | 166 | 180 | 210 |
| 60 | 102 | 153 | 195 | | | | |
| 90 | 165 | | | | | | |
| 105 | 150 | | | | | | |

Table 1: ergodic $\mathcal{ECA}$. Rules are grouped according to minimal equivalence relation [3], since we will prove that they preserve ergodicity, as for $\mathcal{ECA}$.

## 5.1 The case of additive CA

A *topological group* (denoted additively) is a Hausdorff topological space $\mathcal{G}$ endowed with a group operation $+: \mathcal{G} \times \mathcal{G} \to \mathcal{G}$ and satisfying the following properties:

i) $+$ is continuous;

ii) the map $x \to -x$ is continuous.

It is easy to see that in the case of compact spaces i) implies ii).

In this paper we are interested in continuous endomorphisms of compact abelian groups. Consider $\mathcal{G} = \mathbf{Z}_2^{\mathbf{Z}}$, it is a compact abelian group with respect to the usual component-wise addition. Let $G_f: \mathcal{G} \to \mathcal{G}$ be a continuous endomorphism of $\mathcal{G}$. Then for each $i \in \mathbf{Z}$ there exists a finite subset $U_i$ of $\mathbf{Z}$ such that

$$\forall c \in \mathcal{G} \, \forall i \in \mathbf{Z}, \, G_f(c)_i = \sum_{j \in U_i} c_j.$$

If $G_f$ is the global function of a CA then $U_i = \{i - r, \ldots, 0, \ldots, i + r\}$, where $r$ is the radius of the local rule, such CA are called *additive*. Let $\mathcal{ACA}$ be the class of additive CA of fixed radius $r$.

A complete topological group (recall that a compact metric space is always complete) has a left invariant measure if and only if it is locally compact [10].

Moreover this measure is unique up to a positive multiplicative constant. If the group is compact then there exist a unique left invariant measure with total mass 1, we call it the *Haar measure* of the group. The following theorem will be useful in the sequel.

**Theorem 11 [10]** *The Haar measure of a compact group is right invariant and invariant under continuous surjective homomorphisms.*

**Theorem 12 [11]** *One dimensional additive CA are completely mixing.*

Let $p \in \mathbb{N}$ be prime. Recall that the global rule of an additive CA is, also, a group endomorphism i.e. $\forall x, y \in (Z_p)^{\mathbb{Z}}$, $G_f(x + y) = G_f(x) + G_f(y)$; this property will be very useful in the sequel. Let us consider the map $\phi: S^{\mathbb{Z}} \to S^{\mathbb{Z}}$ defined as follows: $\forall x \in S^{\mathbb{Z}} \ \forall i \in \mathbb{Z}$, $\phi(x)_i = \overline{x_i}$.

**Lemma 1** *The map $\phi$ is an isometry.*

**Proof** — The map $\phi$ is continuous since every preimage of an element of the subbase is still an element of the subbase. For all $x, y \in S^{\mathbb{Z}}$ we have:

$$
\begin{aligned}
d(\phi(x), \phi(y)) &= \sum_{i=-\infty}^{+\infty} \frac{\delta(\phi(x)_i, \phi(y)_i)}{2^{|i|}} = \\
&= \sum_{i=-\infty}^{+\infty} \frac{\delta(\overline{x_i}, \overline{y_i})}{2^{|i|}} = \\
&= \sum_{i=-\infty}^{+\infty} \frac{\delta(x_i, y_i)}{2^{|i|}} = d(x, y).
\end{aligned}
$$

●

In this section we consider $S = \{0, 1\}$, $p = 2$ and $r \in \mathbb{N}$. Moreover, for the sake of simplicity, we consider only the case $f(1, \ldots, 1) = 1$, since the other case $f(1, \ldots, 1) = 0$ is perfectly analog.

**Proposition 6** *Let $f \in \mathcal{R}(\{0, 1\}, r)$ be an additive CA. For all $x \in \{0, 1\}^{\mathbb{Z}}$ and for all $t \in \mathbb{N}$ we have:*

$$
G_{\tau_c(f)}^t(x) = \left( \bigoplus_{i=0}^{t-1} G_f^i(\underline{1}) \right) \oplus G_f^t(x).
$$

**Proof** — Let us prove the thesis by induction on $t \in \mathbb{N}$. Let $t = 1$ then

$$
\forall x \in S^{\mathbb{Z}}, \ G_{\tau_c(f)}(x) = G_f^0(\underline{1}) \oplus G_f(x).
$$

Suppose the thesis true for all $n \in \mathbb{N}$ such that $n < t - 1$, then

$$
\begin{aligned}
\forall x \in S^{\mathbb{Z}}, \ G_{\tau_c(f)}^t(x) &= (G_{\tau_c(f)} \circ G_{\tau_c(f)}^{t-1})(x) = \\
&= G_{\tau_c(f)}\left( \left( \bigoplus_{i=0}^{t-2} G_f^i(\underline{1}) \right) \oplus G_f^{t-1}(x) \right) = \\
&= \underline{1} \oplus G_f \left( \bigoplus_{i=1}^{t-2} G_f^i(\underline{1}) \oplus G_f^{t-1}(x) \right) =
\end{aligned}
$$

$$= 1 \oplus \left( \bigoplus_{i=1}^{t} G_f^i(\underline{1}) \right) \oplus G_f^t(x) =$$

$$= \left( \bigoplus_{i=0}^{t} G_f^i(\underline{1}) \right) \oplus G_f^t(x).$$

●

The following corollary is an easy consequence of proposition 6.

**Corollary 2** *Let $f \in \mathcal{R}(\{0,1\}, r)$ be an additive CA. For all $x \in \{0,1\}^{\mathbf{Z}}$ and for all $t \in \mathbf{N}$*

$$G_{\tau_c(f)}^t(x) = \begin{cases} G_f^t(x), & \text{if } t \text{ is even} \\ \underline{1} \oplus G_f^t(x), & \text{otherwise.} \end{cases}$$

**Proposition 7** *Let $\langle \{0,1\}, r, f \rangle$ be an additive CA. Then $\Pi_{G_f} = \Pi_{G_{\tau_c(f)}}$.*

**Proof** — Let $G_f^t(c) = c$, $t > 0$. By corollary 2 we have $G_{\tau_c(f)}^{2t}(c) = c$. We conclude that every periodic point of $G_f$ that have period $t$ is periodic for $G_{\tau_c(f)}$ with period at most $2t$. ●

From proposition 7 and the definition of regular DTDS we have the following.

**Corollary 3** *Let $\mathcal{A} = \langle \{0,1\}, r, f \rangle$ be an additive CA. If $\mathcal{A}$ is regular then $\tau_c(\mathcal{A})$ is regular.*

**Lemma 2** *Let $\langle \{0,1\}, r, f \rangle$ be an additive CA. Then*

$$\forall t \in \mathbf{N}, G_{\tau_c(f)}^{2t+1} = G_f^{2t+1} \circ \phi.$$

**Proof** — We prove the thesis by induction. Let $t = 0$ then

$$G_{\tau_c(f)}(x) = \underline{1} \oplus G_f(x) = G_f(\underline{1}) \oplus G_f(x) = G_f(\underline{1} \oplus x).$$

Suppose that the thesis holds for $t = k - 1$, then

$$\begin{aligned} G_{\tau_c(f)}^{2k+1}(x) &= G_{\tau_c(f)}^2(G_{\tau_c(f)}^{2k-1}(x)) &= G_{\tau_c(f)}^2(G_f^{2k-1}(\underline{1} \oplus x)) = \\ &= G_f^2(G_f^{2k-1}(\underline{1} \oplus x)) &= G_f^{2k+1}(\underline{1} \oplus x). \end{aligned}$$

●

**Proposition 8** *Let $\mathcal{A}$ be an ergodic $\mathcal{ACA}$, then $\tau_c(\mathcal{A})$ is ergodic.*

**Proof** — It suffices to prove that if $G_f$ is ergodic then $G_{\tau_c(f)}$ is ergodic too. Suppose that $G_f$ is ergodic then, by theorem 1 in [11], we have that $\forall n \in \mathbf{N}$ the map $I - G_f^n$ is surjective, that is for all $n \in \mathbf{N}$ and for all $y \in \mathbf{Z}_2^{\mathbf{Z}}$ there exists $x \in \mathbf{Z}_2^{\mathbf{Z}}$ such that $(I - G_f^n)(x) = y$, and, since both $I$ and $G_f$ are additive we have $x - G_f^n(x) = y$. If $n$ is even then, by corollary 2, we have that $I - G_{\tau_c(f)}^n$ is surjective. Now we prove that, if $n$ is odd, for all $\phi(y) \in \mathbf{Z}_2^{\mathbf{Z}}$ the element $\phi(x)$

($x$ is the preimage of $y$ by $G_f$) is the preimage by $I - G^n_{\tau_c(f)}$. Let us consider the following chain of equalities:

$$(I - G^n_{\tau_c(f)})(\phi(x)) = \phi(x) - G^n_{\tau_c(f)}(\phi(x)) \overset{\star}{=} \phi(x) - G^n_f(x) =$$
$$= 1 + x - G^n_f(x) = 1 + y = \phi(y)$$

where $\star$ follows from lemma 2.  •

**Proposition 9** *Let $A = \langle \{0,1\}, r, f \rangle$ be an additive CA. If $A$ is transitive then $\tau_c(A)$ is transitive.*

**Proof** — Since, as for $ACA$, ergodicity implies transitivity and viceversa, by proposition 8 we have that the essential transformed of a transitive $ACA$ is transitive.  •

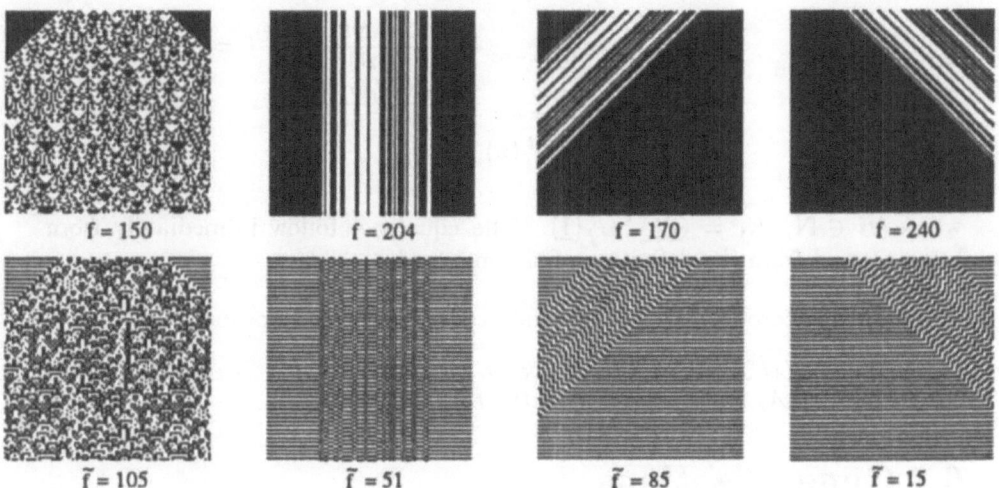

Figure 2: Space-time patterns for additive $ECA$ and their essential conjugates.

From corollary 3 and proposition 9 we obtain the following

**Corollary 4** *Let $A = \langle \{0,1\}, r, f \rangle$ be an additive CA. If $A$ is Devaney (Knudsen) chaotic then $\tau_c(A)$ is Devaney (Knudsen) chaotic.*

In figure 2 we can see an example of the space-time patterns generated by the $ECA$ $f$ and its essential transformed rule $\tilde{f} = \tau_c(f)$. The behavior of rule 204 and $\tau_c(204) = 51$ is respectively fixed point and period 2. As for rules 150 and 105 the space-time diagrams are much more complex. Theorem 1 tells us that rule 150 is Devaney chaotic and then, by corollary 4, rule 105 is Devaney chaotic too.

We are not able to prove that the essential transformed of a strongly transitive $ACA$ is strongly transitive, then we prove a weaker property. Theorem 13 proves that there is a big family of open sets such that the dynamical evolution over them covers the whole phase space.

**Theorem 13** *Let $f \in \mathcal{ACA}$ be strongly transitive then for all non empty open set $A$ we have*

$$\bigcup_{n \in \mathbb{N}} G^n_{\tau_c(f)}(A \cup \phi(A)) = Z_2^{\mathbb{Z}}$$

Finally we introduce an interesting proposition which will be used to prove that $\mathcal{A} = \langle \{0, 1\}, r, f \rangle$ and $\tau_c(\mathcal{A})$ share the same metrical properties.

**Proposition 10** *Let $\langle \{0, 1\}, r, f \rangle$ be an additive CA. Then*

$$\forall x, y \in \{0, 1\}^{\mathbb{Z}} \ \forall t \in \mathbb{N}, \ d(G^t_{\tau_c(f)}(x), G^t_{\tau_c(f)}(y)) = d(G^t_f(x), G^t_f(y)).$$

**Proof** — $\forall x, y \in \{0, 1\}^{\mathbb{Z}} \ \forall t \in \mathbb{N}$,

$$
\begin{aligned}
d(G^t_{\tau_c(f)}(x), G^t_{\tau_c(f)}(y)) &= d\left(K_i \oplus G^t_f(x), K_i \oplus G^t_f(y)\right) = \\
&= \sum_{i=-\infty}^{\infty} \frac{K_i \oplus G^t_f(x)_i \oplus K_i \oplus G^t_f(y)_i}{2^{|i|}} = \\
&= \sum_{i=-\infty}^{\infty} \frac{G^t_f(x)_i \oplus G^t_f(y)_i}{2^{|i|}} = \\
&= d(G^t_f(x), G^t_f(y)).
\end{aligned}
$$

where $\forall i \in \mathbb{N}$, $K_i = \bigoplus_{i=0}^{t-1} G^t_f(\underline{1})$. The equalities follow immediately from lemma 2 and from the definition of the metric $d$. $\qquad \bullet$

The following corollary is an immediate consequence of proposition 10.

**Corollary 5** *Let $\mathcal{A} = \langle \{0, 1\}, r, f \rangle$ be an additive CA. If $\mathcal{A}$ is expansive (sensitive) then $\tau_c(\mathcal{A})$ is expansive (sensitive).*

## 6   Generalizations

We generalize part of the previous results only for additive CA defined over an alphabet of prime cardinality $p$. We prove that $\mathcal{A}$ and $\tau_c(\mathcal{A})$ share both topological (regularity) and metrical (sensitivity,expansivity) properties.

Let $\mathcal{K} = \{\underline{0}, \ldots, \underline{p-1}\}$ and $p$ prime. Here we always assume $f \in \mathcal{ACA}$.

**Lemma 3** *If $G_f(\underline{p-1}) \neq \underline{0}$ then $G_f|_{\mathcal{K}}$ is a permutation.*

**Proof** — Since $\mathcal{K}$ is finite it suffices to prove that $G_f|_{\mathcal{K}}$ is surjective. The hypothesis that $G_f(\underline{p-1}) \neq \underline{0}$ implies that $(a_{-r} + \ldots + a_r)$ is not a divisor of $p$. Since $\mathbb{Z}_p$ is a field, the equation $y = (a_{-r} + \ldots + a_r) \cdot x$ has always a solution in $\mathcal{K}$. Then $G_f|_{\mathcal{K}}$ is surjective. $\qquad \bullet$

**Theorem 14** $\Pi_{\tau_c(f)} = \Pi_f$.

**Proof** — We have to consider two cases:

   i.   $G_f(\underline{p-1}) = \underline{0}$

ii.  $G_f(\underline{p-1}) \neq \underline{0}$.

We prove only the latter since the former is trivial. By lemma 3, $G_f|_\mathcal{K}$ is a permutation then it can be decomposed in a finite number of disjoint cycles. Suppose that the cycle containing the element (p-1) has length $k$ $(1 \leq k \leq p)$. It is easy to see that for all $x \in (\mathbf{Z}_p)^\mathbf{Z}$, $G_f^t(x) = x$ implies that $G_{\tau_c(f)}^{pkt}(x) = x$.

•

**Proposition 11** *For all $x, y \in (Z_p)^\mathbf{Z}$ and for all $t \in \mathbb{N}$ it holds that*

$$d(G_{\tau_c(f)}^t(x), G_{\tau_c(f)}^t(y)) = d(G_f^t(x), G_f^t(y)).$$

**Proof** — For all $t \in \mathbb{N}$ we have

$$
\begin{aligned}
d(G_{\tau_c(f)}^t(x), G_{\tau_c(f)}^t(y)) &= \sum_{i=-\infty}^{+\infty} \frac{\delta(G_{\tau_c(f)}^t(x)_i, G_{\tau_c(f)}^t(y)_i)}{2^{|i|}} = \\
&= \sum_{i=-\infty}^{+\infty} \frac{\delta(K_i + G_f^t(x)_i, K_i + G_f^t(y)_i)}{2^{|i|}} = \\
&\overset{(\star)}{=} \sum_{i=-\infty}^{+\infty} \frac{\delta(G_f^t(x)_i, G_f^t(y)_i)}{2^{|i|}} = \\
&= d(G_f(x), G_f(y))
\end{aligned}
$$

where $\forall i \in \mathbb{N}$, $K_i = (\sum_{i=0}^{t-1} G_f(\underline{p-1}))_i \in \mathbf{Z}_p$. Equality $(\star)$ holds since $\mathbf{Z}_p$ is a group (cancellation laws).

•

From proposition 11 we have the generalized version of corollary 5.

**Corollary 6** *If $G_f$ is expansive (sensitive) then $G_{\tau_c(f)}$ is expansive (sensitive).*

# 7  Conclusions

We have reviewed some of the most popular definitions of chaos noting that they are insufficient to discriminate the chaotic behavior of a biinfinite shift from the one of a system that is topologically conjugate to a one-sided shift. This latter kind of chaotich systems seem to fit better the intuitive concept of chaos even from a pictorial point of view. A possible solution to this problem is the introduction of CF-chaotic systems. This definition is not empty since $\mathcal{LRCA}$ are CF-chaotic. We also study the correlations with the other definitions. In particular the problem of establishing if CF-chaos implies Devaney's one. We proved that the definitions of Devaney's and Knudsen's chaos are invariant if we restrict the rule space to $T_e(\mathcal{ACA})$. The authors suspect that the same result holds also for CF-chaos but this problem is still open.

# References

[1] J. Banks, J. Brooks, G. Cairns, G. Davis, and P. Stacey. On Devaney's definition of chaos. *Amer. Math. Mountly*, pages 332–334, 1991.

[2] G. Cattaneo, E. Formenti, G. Manzini, and L. Margara. On ergodic linear cellular automata over $z_m$. Submitted to Stacs, 1996.

[3] G. Cattaneo, E. Formenti, L. Margara, and G. Mauri. Transformations of the one dimensional cellular automata rule space. Technical Report 171-96, Università degli Studi di Milano, Dep. of Computer Science, via Comelico 39, Milan, Italy, 1996. Submitted to Parallel Computing, special issue on cellular automata.

[4] G. Cattaneo, E. Formenti, and C. Quaranta Vogliotti. A new algorithmic procedure for the $\Phi \overset{?}{=} \{\underline{0}\}$ problem for cellular automata. Preprint, 1994.

[5] B. Codenotti and L. Margara. Transitive cellular automata are sensitive. *American Mathematical Monthly*, 103:??, 1996.

[6] R. L. Devaney. *Introduction to chaotic dynamical systems*. Addison-Wesley, second edition, 1989.

[7] F. Fagnani and L. Margara. Expansivity, permutivity, and chaos for cellular automata. Submitted toTheoretical Computer Science, 1996.

[8] M. Finelli, G. Manzini, and L. Margara. Lyapunov exponents Vs expansivity and sensitivity in cellular automata. Submitted to ACRI '96, 1996.

[9] G. A. Hedlund. Endomorphism and automorphism of the shift dynamical system. *Mathematical System Theory*, 3:320–375, 1969.

[10] R. Mañé. *Ergodic theory and differentiable dynamics*, volume 8 of *Modern surveys in mathematics*. Springer-Verlag, Berlin, 1987.

[11] M. Shirvani and T. D. Rogers. Ergodic endomorphisms of compact abelian groups. *Communications in Mathematical Physics*, 118:401–410, 1988.

[12] M. Shirvani and T. D. Rogers. On ergodic one-dimensional cellular automata. *Communications in Mathematical Physics*, 136:599–605, 1991.

[13] K. Sutner. Fischer automata and the *edge of chaos*. Preprint, 1994.

# Lyapunov Exponents Vs Expansivity and Sensitivity in Cellular Automata

Michele Finelli

Dipartimento Scienze dell'Informazione, Università di Bologna
Bologna, Italy

Giovanni Manzini

Dipartimento di Scienze e Tecnologie Avanzate, Università di Torino
Alessandria, Italy.

Luciano Margara

Dipartimento Scienze dell'Informazione, Università di Bologna
Bologna, Italy

### Abstract

We establish a connection between the theory of Lyapunov exponents and the properties of expansivity and sensitivity to initial conditions for a particular class of discrete time dynamical systems: the cellular automata. The main result of this paper is the proof that all the expansive cellular automata have positive Lyapunov exponents for almost all the phase space configurations. A rather surprising corollary of this result is that there are no expansive CA in any dimension higher than 1. Finally, we prove some interesting properties of expansive and sensitive cellular automata.

## 1 Introduction

The notion of chaos is very appealing, and it has intrigued many scientists (see [1, 2, 5, 10, 14] for some works on the properties that characterize a chaotic process). There are simple deterministic dynamical systems that exhibit unpredictable behavior. Though counterintuitive, this fact has a very clear explanation. The lack of infinite precision causes a loss of information which is dramatic for some processes which quickly loose their deterministic nature to assume a non deterministic (unpredictable) one.

A chaotic phenomenon can indeed be viewed as a deterministic one, in the presence of infinite precision, and as a nondeterministic one, in the presence of finite precision constraints (see Figure 1). Thus one should look at chaotic processes as at processes merged into time, space, and precision bounds, which are the key resources in the science of computing.

A nice way in which one can analyze this finite/infinite dichotomy is by using cellular automata models (CA). Consider the 1-dimensional CA $(X, \sigma)$, where $X = \{0, 1\}^{\mathbf{Z}}$ and $\sigma$ is the shift map on $X$.

In order to completely describe the elements of $X$, we need to operate on sequences of binary digits of infinite length. Assume for a moment that this is

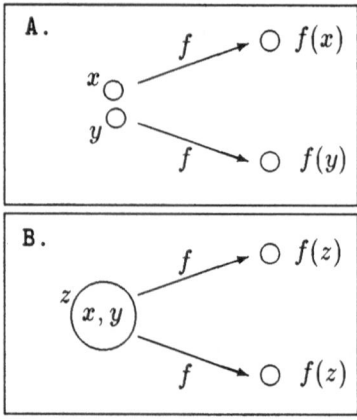

Figure 1: (A) $f$ is deterministic. In the case of finite precision (B) $x$ and $y$ are not distinguishable and the map $f$ looses its deterministic nature.

possible. Then the shift map is completely predictable, i.e., one can completely describe $\sigma^n(x)$, for any $x \in X$ and for any integer $n$.

In practice only finite objects can be computationally manipulated. Let $x \in X$. Assume we know a portion of $x$ of length $n$ (the portion between the two vertical lines in Figure 2). One can easily verify that $\sigma^n(x)$ completely depends on the unknown portion of $x$. In other words, if we have finite precision, the shift map becomes unpredictable, as a consequence of the combination of the finite precision representation of $x$ and the *sensitivity* of $\sigma$.

In the case of discrete time dynamical systems (DTDS) defined on a metric space, many definitions of chaos are based on the notion of sensitivity (see for example [5, 6, 9]).

We now recall the definition of sensitivity to initial conditions for a DTDS $(X, F)$. Here, we assume that $X$ is equipped with a distance $d$ and that the map $F$ is continuous on $X$ according to the metric topology induced by $d$.

**Definition 1 (Sensitivity)** *A DTDS $(X, F)$ is sensitive to initial conditions if and only if there exists $\delta > 0$ such that for any $x \in X$ and for any neighborhood $N(x)$ of $x$, there is a point $y \in N(x)$ and a natural number $n$, such that $d(F^n(x), F^n(y)) > \delta$. $\delta$ is called the sensitivity constant.*

Intuitively, a map is sensitive to initial conditions, or simply sensitive, if there exist points arbitrarily close to $x$ which eventually separate from $x$ by at least $\delta$ under iteration of $F$. We emphasize that not all points near $x$ need eventually separate from $x$, but there must be at least one such point in every neighborhood of $x$. If a map possesses sensitive dependence on initial conditions, then for all practical purposes, the dynamics of the map defies numerical approximation. Small errors in computation which are introduced by round-off may become magnified upon iteration. The results of numerical computation of an orbit, no matter how accurate, may be completely different from the real orbit.

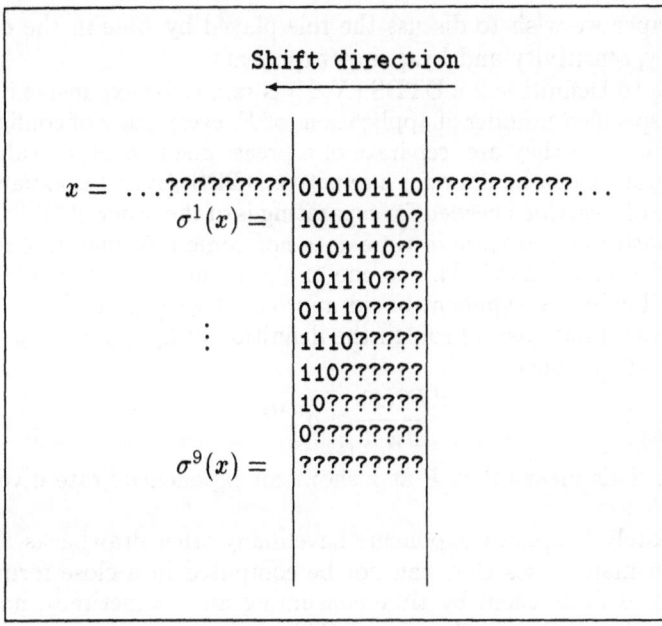

Figure 2: Finite precision combined with sensitivity to initial conditions causes unpredictability after a few iterations ($x$ represents the state of the CA at time step 0, and $\sigma^i(x)$ the state at time step $i$).

A stronger property is expansiveness. Expansiveness differs from sensitivity in that all nearby points eventually separate by at least $\delta$. It is easy to verify that expansive CA are sensitive to initial conditions.

**Definition 2 (Expansivity)** *A DTDS $(X, F)$ is expansive if and only if there exist $\epsilon, \delta > 0$ such that for every $x, y \in X$*

$$d(x, y) \leq \epsilon \implies \exists n \geq 0 \text{ such that } d(F^n(x), F^n(y)) > \delta.$$

*The value $\delta$ is called the expansivity constant.*

In the case of differentiable spaces there is another parameter which is often used for detecting chaotic behaviors: the Lyapunov exponents. We define them for a map $F$, $F : I \to I$, where $I$ is a real interval.

**Definition 3 (Lyapunov Exponents)** *Let $(I, F)$ be a DTDS. The Lyapunov exponent $\lambda(x)$ of $x \in X$ is defined by*

$$\lambda(x) = \lim_{n \to \infty} \frac{1}{n} \log \left( \frac{d \, F^n(x)}{d \, x} \right)$$

Lyapunov exponents can be easily generalized to higher dimensions. Usually, a DTDS $(X, F)$ is said to be chaotic at $x \in X$ if and only if $\lambda(x) > 0$.

In this paper we wish to discuss the role played by *time* in the definitions of expansivity, sensitivity and Lyapunov exponents.

According to Definition 2 a DTDS $(X, F)$ is said to be expansive if and only if after an unspecified number of applications of $F$, every pairs of configurations, no matter how close they are, separate of a preassigned constant value $\delta$. The same consideration can be done for a sensitive DTDS. $(X, F)$ is expansive even if the number of iterations needed for separating is of the order of $10^{100}$. In other words, a quantitative measure of time does not come into play in determining the expansivity of a DTDS. This is one of the main criticism made by those who prefer a Lyapunov exponent based approach for defining chaos.

Time plays a fundamental role in the definition of Lyapunov exponents. In fact, if $\lambda(x) > 0$, we have

$$\frac{d\,F^n(x)}{d\,x} \simeq \alpha^{n\lambda(x)}$$

where $\alpha > 1$. This means that $F$ at $x$ shows an exponential rate divergence in time.

Unfortunately Lyapunov exponents have many other drawbacks. The main one is that in many cases they can not be computed in a close form and one needs to approximate them by time consuming and, sometimes, not reliable computer simulations (as in the case of CA).

In this paper we prove that the criticisms made by the supporters of the Lyapunov exponents to the expansivity property are not well founded. In fact, we show that every expansive CA must have almost all the Lyapunov exponents uniformly bounded away from zero by a constant $\delta$ which only depends on the CA we consider. In other words, expansivity implies positive Lyapunov exponents. Note that the task of verifying expansivity appears to be simpler than the computation of the Lyapunov exponents. In [7], the authors define a large class of expansive CA which contains additive and non additive ones.

Our result is based on some properties of expansive Lipschitz functions which hold for any compact metric space. As a byproduct of our analysis we obtain several results on expansive and sensitive Lipschitz functions which are of independent interest.

The main results of this paper can be summarized as follows.

1. Let $(X, F)$ be any expansive CA with expansivity constant $\delta$. Let $x, y \in X$ be any pair of distinct configurations whose distance is $\epsilon$. The number of iterations needed by F for separating $x$ and $y$ by at least $\delta$ does not depend on $x$ and $y$ and is of the order of $\log(\frac{\delta}{\epsilon})$ (Theorem 4.1).

2. Every expansive CA $(X, F)$ has positive (uniformly bounded away from zero) Lyapunov exponents over a set $Y \subseteq X$ of configurations of full measure. A slightly weaker result holds also for those configurations belonging to $X \setminus Y$ (Theorem 5.1).

3. Expansive CA exist only in 1 dimension (Theorem 5.2).

In addition, we highlight the different behavior of expansive and sensitive CA for what concerns the speed at which perturbations propagate.

The rest of this paper is organized as follows. In Section 2 we give some definitions and preliminary material. In Section 3 we introduce the notion of

Lyapunov exponents for CA and we recall some known results concerning CA and Lyapunov exponents. In Section 4 we prove some properties of expansive functions over a compact metric space, and in Section 5 we show the relationship between expansivity and Lyapunov exponents in CA. In Section 6 we discuss sensitive CA, and in Section 7 we draw the conclusions of the paper.

## 2    Definitions and notations

Let $\mathcal{A} = \{0, 1, \ldots, p - 1\}$ be a finite alphabet of cardinality $p \geq 2$. Let $f$, $f \colon \mathcal{A}^s \to \mathcal{A}$, $s \geq 1$, be any map. We say that $s$ is the size of the domain of $f$, or simply the size of $f$.

A $D$-dimensional CA based on a *local rule* $f$ of size $s$ is a pair $(\mathcal{A}^{\mathbf{Z}^D}, F)$, where

$$\mathcal{A}^{\mathbf{Z}^D} = \{c \mid c : \mathbf{Z}^D \to \mathcal{A}\}$$

is the *space of configurations* and $F$, $F : \mathcal{A}^{\mathbf{Z}^D} \to \mathcal{A}^{\mathbf{Z}^D}$, is the *global transition map* defined as follows. For every $c \in \mathcal{A}^{\mathbf{Z}^D}$ and for every $\vec{v} \in \mathbf{Z}^D$

$$[F(c)](\vec{v}) = f\left(c(\vec{v} + nb(1)), \ldots, c(\vec{v} + nb(s))\right),$$

where $nb$, $nb : \{1, \ldots, s\} \to \mathbf{Z}^D$, is the *neighborhood structure map*.

In the case of 1-dimensional CA, we use the following simplified notation. Let $f, f : \mathcal{A}^{2k+1} \to \mathcal{A}$, be any map. A 1-dimensional CA based on the local rule $f$ is a pair $(\mathcal{A}^{\mathbf{Z}}, F)$, where $\mathcal{A}^{\mathbf{Z}}$ is the space of configurations and $F$, $F : \mathcal{A}^{\mathbf{Z}} \to \mathcal{A}^{\mathbf{Z}}$, is defined by

$$[F(c)](i) = f(c(i - k), \ldots, c(i + k)), \quad c \in \mathcal{A}^{\mathbf{Z}}, \ i \in \mathbf{Z}.$$

We say that $k$ is the *radius* of $f$. Note that, even if $f$ must depend on at least one between $x_{-k}$ and $x_k$, in general $f$ does not depend on all the $2k + 1$ variables $x_{-k}, \ldots, x_k$.

In order to specialize the notion of expansivity to the special case of $D$-dimensional CA we introduce a distance over the space of the configurations. Let $\Delta \colon \mathcal{A} \times \mathcal{A} \to \{0, 1\}$ be such that

$$\Delta(i, j) = \begin{cases} 0, & \text{if } i = j, \\ 1, & \text{if } i \neq j. \end{cases}$$

Given $a, b \in \mathcal{A}^{\mathbf{Z}^D}$ the Tychonoff distance $d(a, b)$ is defined by

$$d(a, b) = \sum_{\vec{v} \in \mathbf{Z}^D} \frac{\Delta(a(\vec{v}), b(\vec{v}))}{2^{\max(\vec{v})}}, \tag{1}$$

where $\max(\vec{v})$ is the maximum of the absolute value of the components of $\vec{v}$. It is easy to verify that $d$ is a metric on $\mathcal{A}^{\mathbf{Z}^D}$ and that the metric topology induced by $d$ coincides with the product topology induced by the discrete topology of $\mathcal{A}$. With this topology, $\mathcal{A}^{\mathbf{Z}^D}$ is a compact and totally disconnected space and $F$ is a (uniformly) continuous map.

Throughout the paper, $F(c)$ will denote the result of the application of the map $F$ to the configuration $c$, $c(\vec{v})$ will denote the value of the entry with coordinates $\vec{v}$ of the configuration $c$, and $\vec{v}_i$ will denote the i-th component of the vector $\vec{v}$. We recursively define $F^n(c)$ by $F^n(c) = F(F^{n-1}(c))$, where $F^0(c) = c$.

# 3 Lyapunov exponents for CA

The notion of Lyapunov exponents given in Definition 3 can be applied only to differentiable spaces. Since $\mathcal{A}^{\mathbf{Z}}$ is not a differentiable space, we need to use an *ad-hoc* definition. In this section we recall the definition of Lyapunov exponents for the special case of 1-dimensional CA given in [13]. There, the authors introduce in a reasonable way quantities analogous to Lyapunov exponents of smooth dynamical systems which describe the *local* instability of orbits in CA.

For every $x \in \mathcal{A}^{\mathbf{Z}}$ and $s \geq 0$ we set

$$
\begin{aligned}
W_s^-(x) &= \left\{ y \in \mathcal{A}^{\mathbf{Z}} : y(i) = x(i) \text{ for all } i \geq s \right\}, \\
W_s^+(x) &= \left\{ y \in \mathcal{A}^{\mathbf{Z}} : y(i) = x(i) \text{ for all } i \leq s \right\}.
\end{aligned}
$$

We have that $W_i^+(x) \subset W_{i+1}^+(x)$ and $W_i^-(x) \subset W_{i+1}^-(x)$. For every $n \geq 0$ we define

$$
\begin{aligned}
\tilde{\Lambda}_n^-(x) &= \min \left\{ s \geq 0 : F^n(W_0^+(x)) \subset W_s^-(x) \right\}, \\
\tilde{\Lambda}_n^+(x) &= \min \left\{ s \geq 0 : F^n(W_0^-(x)) \subset W_s^+(x) \right\}.
\end{aligned}
$$

Intuitively, for the CA defined by $F$ the value $\tilde{\Lambda}_n^+(x)$ $[\tilde{\Lambda}_n^-(x)]$ measures how far a perturbation front moves right [left] in time $n$ if the front is initially located at $i = 0$. Finally, we consider the following shift invariant quantities

$$
\Lambda_n^-(x) = \max_{j \in \mathbf{Z}} \tilde{\Lambda}_n^-(\sigma^j(x)), \qquad \Lambda_n^+(x) = \max_{j \in \mathbf{Z}} \tilde{\Lambda}_n^+(\sigma^j(x)).
$$

The values $\lambda^+(x)$ and $\lambda^-(x)$ defined by

$$
\lambda^+(x) = \lim_{n \to \infty} \frac{1}{n} \Lambda_n^+(x) \qquad \lambda^-(x) = \lim_{n \to \infty} \frac{1}{n} \Lambda_n^-(x) \tag{2}
$$

are called respectively the right and left Lyapunov exponents of the CA $F$ for the configuration $x$. The limits in (2) do not necessarily exist for all $x \in \mathcal{A}^{\mathbf{Z}}$. However, the following result holds.

**Theorem 3.1** *[13] For any $\sigma$-invariant and $F$-invariant measure $\mu$ defined on $\mathcal{A}^{\mathbf{Z}}$, there exists a set $Y \subseteq X$ of full measure ($\mu(Y) = 1$) such that for every $x \in Y$ the limits (2) exist.*

# 4   Some properties of expansive functions

In this section we prove some properties of expansive functions over a compact metric space. In particular, we consider the case in which we are given a function $F: X \to X$ such that

$$\exists \lambda > 0: \quad \forall x, y \in X, \quad d(F(x), F(y)) \le \lambda d(x, y) \tag{3}$$

Using calculus terminology, if (3) holds we say that $F$ is a Lipschitz function with parameter $\lambda$. The reason for which we are interested in Lipschitz functions is that the global transition map $F$ associated to a CA always satisfies (3). In this case, the parameter $\lambda$ can be easily obtained from the radius of the local rule.

If $d(x, x')$ is small, eq. (3) tells us that the distance $d(F^n(x), F^n(x'))$ cannot grow arbitrarily fast. In this section we show that, if $F$ is expansive, there is also a lower bound on how fast the distance $d(F^n(x), F^n(x'))$ can grow (Theorem 4.1). This result have some remarkable consequences for CA that will be discussed in Section 5.

**Lemma 1** Let $(X, d)$ be a compact metric space, and $F: X \to X$ an expansive Lipschitz function. Then there exist $\epsilon > 0$ and an integer $n$ such that

$$\forall x, y \in X \quad d(x, y) \le \epsilon \implies \exists k \le n \text{ such that } d(F^k(x), F^k(y)) > 2d(x, y).$$

**Proof.** Assume that $F$ is expansive with parameters $\delta$ and $\epsilon'$, and let $\lambda$ denote the Lipschitz constant of the function $F$. Note that, since $F$ is expansive, we must have $\lambda > 1$. We prove the theorem for $\epsilon = \min(\delta/6, \epsilon'/2)$.

Assume by contradiction that

$$\forall n \, \exists x_n, y_n: \quad d(x_n, y_n) \le \epsilon \text{ and } d(F^k(x_n), F^k(y_n)) \le 2d(x_n, y_n) \quad \forall k \le n.$$

Since $X$ is a compact space, we can build two sequences $x_i, y_i$ such that

$$d(x_i, y_i) \le \epsilon \le \delta/6, \quad \lim_{i \to \infty} x_i = \tilde{x}, \quad \lim_{i \to \infty} y_i = \tilde{y}, \tag{4}$$

and

$$d(F^k(x_i), F^k(y_i)) \le 2d(x_i, y_i) \quad \forall k \le i. \tag{5}$$

Moreover, we can assume that

$$d(x_i, \tilde{x}) < \frac{\delta}{3}\lambda^{-i}, \quad d(y_i, \tilde{y}) < \frac{\delta}{3}\lambda^{-i}. \tag{6}$$

By the triangle inequality, we have

$$
\begin{aligned}
d(\tilde{x}, \tilde{y}) &\le d(\tilde{x}, x_i) + d(x_i, y_i) + d(y_i, \tilde{y}) \\
&\le \frac{2\delta}{3}\lambda^{-i} + \epsilon'/2.
\end{aligned}
$$

Since this is true for all $i$, we have that $d(\tilde{x}, \tilde{y}) \le \epsilon'$. Hence, for the expansivity of $F$, there exists $m$ such that $d(F^m(\tilde{x}), F^m(\tilde{y})) > \delta$. Using again the triangle inequality, we get

$$
\begin{aligned}
d(F^m(\tilde{x}), F^m(\tilde{y})) &\le d(F^m(\tilde{x}), F^m(x_m)) + d(F^m(x_m), F^m(y_m)) + \\
&\quad d(F^m(y_m), F^m(\tilde{y})),
\end{aligned}
$$

which, using (6) and (4), yields

$$
\begin{aligned}
d(F^m(x_m), F^m(y_m)) \;&\geq\; d(F^m(\tilde{x}), F^m(\tilde{y})) - d(F^m(\tilde{x}), F^m(x_m)) - \\
&\qquad d(F^m(\tilde{y}), F^m(y_m)), \\
&>\; \delta - \lambda^m d(\tilde{x}, x_m) - \lambda^m d(\tilde{y}, y_m), \\
&>\; \frac{\delta}{3}, \\
&\geq\; 2d(x_m, y_m).
\end{aligned}
$$

Hence, $d(F^m(x_m), F^m(y_m)) > 2d(x_m, y_m)$ which is impossible since it contradicts the hypothesis (5). □

The previous lemma establishes that, if $d(x, y)$ is small enough, by applying $F$ at most $n$ times we can double the original distance. The followinq theorem shows that there exists $\tilde{n}$ such that, by applying $F$ $\tilde{n}$ times we double the original distance for all pairs $x, y$ with $d(x, y) \leq \epsilon'$.

**Theorem 4.1** *Let $(X, d)$ be a compact metric space, and $F: X \to X$ be an expansive Lipschitz function. Then there exist $\epsilon' > 0$ and $\tilde{n} \geq 1$ such that*

$$\forall x, y \in X, \quad d(x, y) \leq \epsilon \implies d(F^{\tilde{n}}(x), F^{\tilde{n}}(y)) > 2d(x, y).$$

**Proof.** Let $\lambda$ and $\delta$ denote respectively the Lipschitz constant and the expansivity constant of $F$. By Lemma 1 we know that there exist $\epsilon, n$ such that

$$d(x, y) \leq \epsilon \implies \exists k \leq n \text{ such that } d(F^k(x), F^k(y)) > 2d(x, y).$$

Let $M = \lceil n \log \lambda \rceil + 1$. We prove the theorem for $\epsilon' = \epsilon \lambda^{-nM}$, and $\tilde{n} = n(M-1) = n \lceil n \log \lambda \rceil$.

Assume by contradiction that $\exists x, y$ such that $d(x, y) < \epsilon'$ and

$$d(F^{\tilde{n}}(x), F^{\tilde{n}}(y)) \leq 2d(x, y). \tag{7}$$

To prove that this is impossible, we first show that $\exists \tilde{k}$, $\tilde{n} < \tilde{k} \leq nM$, such that

$$d\left(F^{\tilde{k}}(x), F^{\tilde{k}}(y)\right) > 2^M d(x, y). \tag{8}$$

For $k \leq Mn$ we have $d(F^k(x), F^k(y)) \leq \lambda^k d(x, y) \leq \epsilon$, hence, all pairs $(F^k(x), F^k(y))$ satisfy the hypothesis of Lemma 1. By applying the lemma $M$ times, we have that there exists $k_0 \leq nM$ such that

$$d(F^{k_0}(x), F^{k_0}(y)) > 2^M d(x, y).$$

If $k_0 > \tilde{n}$, we set $\tilde{k} = k_0$ and we are done. Otherwise, we apply Lemma 1 to $(F^{k_0}(x), F^{k_0}(y))$, and we get $k_1$, $(k_0 < k_1 \leq k_0 + n)$, such that

$$d(F^{k_1}(x), F^{k_1}(y)) > 2d(F^{k_0}(x), F^{k_0}(y)) > 2^{M+1} d(x, y).$$

We continue in this way until we get an integer $k_r$ such that $d(F^{k_r}(x), F^{k_r}(y)) > 2^{M+r} d(x, y)$. The value $\tilde{k} = k_r$ clearly satisfies (8).

Let $m = \tilde{k} - \tilde{n}$, $\tilde{x} = F^{\tilde{n}}(x)$, and $\tilde{y} = F^{\tilde{n}}(y)$. Since $F^{\tilde{k}}(x) = F^m(\tilde{x})$, combining (8) with (7) we get

$$d(F^m(\tilde{x}), F^m(\tilde{y})) > 2^M d(x, y) \geq 2^{M-1} d(\tilde{x}, \tilde{y}) \geq \lambda^n d(\tilde{x}, \tilde{y})$$

which is impossible since by (3) we have

$$d(F^m(\tilde{x}), F^m(\tilde{y})) \leq \lambda^m d(\tilde{x}, \tilde{y}),$$

and $m \leq n$. □

## 5 Expansivity and Lyapunov exponents in Cellular Automata

In this section we prove some results concerning expansive CA. Our results follow from the properties of expansive Lipschitz functions established in the previous section. The fundamental observation is that for any expansive CA Lemma 1 and Theorem 4.1 establish that there is a lower bound to the speed at which "perturbations" propagate in the configuration space. A first consequence of this fact, is that expansive CA have positive Lyapunov exponents. In order to prove this result we need a preliminary lemma.

**Lemma 2** Let $(\mathcal{A}^{\mathbf{Z}}, F)$ be an expansive 1-dimensional CA. For every $x \in (\mathcal{A}^{\mathbf{Z}})$, let $\Lambda_n^+(x), \Lambda_n^-(x)$ defined as in Section 3. Then, there exists a constant $c > 0$ such that

$$\limsup_{n \to \infty} \frac{1}{n} \Lambda_n^+(x) \geq c, \quad and \quad \limsup_{n \to \infty} \frac{1}{n} \Lambda_n^-(x) \geq c.$$

**Proof.** Since $F$ is expansive, by Lemma 1 we can find $\epsilon, m$ such that $\forall y, z$

$$d(y, z) \leq \epsilon \implies \exists k \leq m \text{ such that } d(F^k(y), F^k(z)) > 2d(y, z). \tag{9}$$

We prove the lemma by showing that there exists an infinite set of integers $\{n_j\}_{j \in \mathbf{N}_+}$ such that

$$\frac{1}{n_j} \Lambda_{n_j}^-(x) \geq \frac{1}{m},$$

(the proof for $\Lambda_n^+(x)$ is similar).

Let $s$ be an integer such that $2^{-s} < \epsilon$, and let $x'$ be a configuration such that $x'(i) = x(i)$ for $i \neq 0$ and $x'(0) \neq x(0)$. For any integer $j > 0$, we have $d(\sigma^{j+s}(x), \sigma^{j+s}(x')) = 2^{-(j+s)}$. By applying eq. (9) $j$ times we have that there exists $n_j \leq jm$ such that

$$d(F^{n_j}(\sigma^{j+s}(x)), F^{n_j}(\sigma^{j+s}(x'))) > 2^j d(\sigma^{j+s}(x), \sigma^{j+s}(x')) \tag{10}$$

$$= 2^{-s}.$$

This means that while $\sigma^{j+s}(x)$ and $\sigma^{j+s}(x')$ differ only at position $j + s$, $F^{n_j}(\sigma^{j+s}(x))$ and $F^{n_j}(\sigma^{j+s}(x'))$ must differ at positions $\leq s$. Since $\Lambda_{n_j}^-(x)$ measures how far a perturbation can move right in $n_j$ steps, we have

$$\frac{1}{n_j}\Lambda_{n_j}^-(x) \geq \frac{j}{n_j} \geq \frac{1}{m}$$

as claimed. To complete the proof we must show that the set $\{n_j\}_{j \in \mathbf{N}_+}$ contains an infinite number of elements. To prove this we simply observe that, since $F$ is a Lipschitz function, eq. (10) implies that $n_j > j/\log \lambda$. $\quad\square$

We are now ready to prove the main result of the paper.

**Theorem 5.1** *Let $(\mathcal{A}^{\mathbf{Z}}, F)$ be an expansive 1-dimensional CA, and let $Y$ denote the subset of $\mathcal{A}^{\mathbf{Z}}$ for which the right and left Lyapunov exponents ($\lambda^+$ and $\lambda^-$) exist. Then, there exists a constant $c > 0$ such that for all $x \in Y$*

$$\lambda^+(x) \geq c, \quad and \quad \lambda^-(x) \geq c.$$

*Moreover, for any $\sigma$-invariant and $F$-invariant measure $\mu$ there exists a $\mu$-measurable set $Z_\mu$ such that $Z_\mu \subseteq Y$ and $\mu(Z_\mu) = 1$.*

**Proof.** Since $\lambda^+(x)$ and $\lambda^-(x)$ are defined as

$$\lambda^+(x) = \lim_{n \to \infty} \frac{1}{n}\Lambda_n^+(x), \qquad \lambda^-(x) = \lim_{n \to \infty} \frac{1}{n}\Lambda_n^-(x)$$

if the limits exist they cannot be smaller than the constant $c$ given by Lemma 2. The second part of the theorem follows directly by Theorem 3.1. $\quad\square$

Note that, if $F$ is expansive it is also surjective [7]. Hence, the Haar measure is $F$-invarianat and $\sigma$-invariant [12] and therefore satisfies the hypothesis of Theorem 5.1.

A surprising consequence of the fact that in expansive CA perturbations propagate with a speed which is uniformly bounded away from zero, is that there are no expansive CA with dimension $D \geq 2$.

**Theorem 5.2** *Let $\mathcal{A}$ denote any finite alphabet. For the topology induced by the Tychonoff distance (1), if $D \geq 2$ there are no expansive CA $(\mathcal{A}^{\mathbf{Z}^D}, F)$.*

**Proof.** Let $p = |\mathcal{A}|$. We prove the result for $D = 2$, but the same reasoning can be applied for all $D \geq 2$. For any positive integer $m$, we define the set $Q_m \subset \mathbf{Z}^2$ as follows:

$$Q_m = \{(v_1, v_2) \in \mathbf{Z}^2 | \max(|v_1|, |v_2|) \leq m\}.$$

Clearly, $|Q_m| = (2m + 1)^2$. In addition, elementary calculus shows that for each pair $x, y \in \mathcal{A}^{\mathbf{Z}^2}$, we have

$$x(\vec{v}) = y(\vec{v}) \quad \forall \vec{v} \in Q_m \quad \Longrightarrow \quad d(x, y) \leq t_m = 8\frac{(m + 2)}{2^m}.$$

In other words, if two configurations differ only outside $Q_m$ their distance is bounded by the quantity $t_m$.

Now assume $F$ is expansive, and let $\epsilon, n$ denote the values given by Lemma 1. Let $s$ be the smallest integer such that $t_s \leq \epsilon$. This choice guarantees that, if $x, y$ differ only outside $Q_s$, then $d(x, y) \leq \epsilon$ and we can apply Lemma 1.

Let $r > s$, and $z$ be any configuration in $\mathcal{A}^{\mathbf{Z}^2}$. We define $B_{z,r}$ as the set of configurations which coincide with $z$ outside $Q_r$. That is,

$$B_{z,r} = \{x \in \mathcal{A}^{\mathbf{Z}^2} \mid x(\vec{v}) = z(\vec{v}) \, \forall \vec{v} \notin Q_r\}.$$

Clearly, $|B_{z,r}| = p^{|Q_r|} = p^{(2r+1)^2}$. Moreover, $x, y \in B_{z,r}$ implies $d(x, y) \geq 2^{-r}$.

Consider now any pair $x, y \in B_{z,r}$, and let $M$ be an integer such that $2^{M-r} > t_s$. By applying Lemma 1 $M$ times, we can find an integer $k \leq nM$ such that

$$d\big(F^k(x), F^k(y)\big) > t_s. \tag{11}$$

This means that the two configurations $F^k(x)$ and $F^k(y)$ must differ inside $Q_s$. We prove that $F$ cannot be expansive by showing that this is not possible for all pairs $x, y \in B_{z,r}$.

For any configuration $x$, let $x(Q_s)$ denote the set of values assumed by $x$ inside $Q_s$. For $x \in B_{z,r}$ we define the *orbit* of $x$ as the set

$$O_x = \bigcup_{i=0}^{nM} \big[F^i(x)\big](Q_s).$$

The orbit $O_x$ represents the values assumed inside $Q_s$ by the sequence $x, F(x)$, $\ldots, F^{nM}(x)$. If $F$ is expansive, all orbits $O_x$, for $x \in B_{z,r}$, must be distinct (otherwise (11) is violated). We prove that this is impossible by showing that, for $r$ sufficiently large, the number of possible orbits is less than $|B_{z,r}|$.

The number of distinct orbits is given by $(p^{|Q_s|})^{nM+1}$. Since $M$ must be such that $2^{M-r} > t_s$, we can take $M \approx (r-s) + \log s$. Thus, asymptotically the number of orbits is given by $\approx p^{4s^2 n(r-s)}$. Since $n$ and $s$ are constants, we have that for $r$ large enough this is less than $|B_{z,r}| = p^{(2r+1)^2}$.

This completes the proof. $\square$

# 6 Expansivity vs sensitivity in Cellular Automata

In this section we study the properties of sensitive functions over a compact metric space. Again, we assume that the function $F: X \to X$ satisfies the Lipschitz property (3). To simplify our proofs, we rewrite the definition of sensitive function given in section 2 using the $\epsilon, \delta$ notation. Thus, we say that the function $F$ is sensitive if there exists $\delta > 0$ such that

$$\forall x \forall \epsilon \, \exists y: \, d(x, y) \leq \epsilon \text{ and } d(F^n(x), F^n(y)) > \delta \text{ for some } n \geq 0.$$

Since the definitions of sensitivity and expansivity are somewhat similar, it is natural to ask whether a result similar to Lemma 1 holds for sensitive functions.

```
···*****11111000000000000000*****···
···*****11111100000000000000*****···
···*****11111110000000000000*****···
···*****11111111000000000000*****···
···*****11111111100000000000*****···
···*****11111111110000000000*****···
```

Figure 3: The evolution of configuration $y$. Since $f(\alpha, 1, 1) = 1$, and $f(0, 0, \alpha) = 0$ the positions marked with $*$ do not affect the central section.

The answer is no. The following example shows a sensitive Lipschitz function $F$ such that $\forall n$ we can find $x_n \in X$ and a value $\epsilon_n$ such that

$$d(x_n, y) \leq \epsilon_n \implies d\big(F^k(x_n), F^k(y)\big) \leq 2d(x_n, y) \quad \text{for } 0 \leq k \leq n. \quad (12)$$

In other words, we show that there cannot be an upper bound to the number of iterations required for doubling the distance between two arbitrarily close configurations.

**Example 1** We consider the one-dimensional CA over the alphabet $\mathcal{A} = \{0, 1, 2\}$ defined by the following local rule (here $\alpha, \beta$ denote any symbol in $\mathcal{A}$):

$$f(\alpha, 1, 2) = 2, \qquad f(\alpha, 1, 1) = 1, \qquad f(\alpha, 1, 0) = 1, \qquad (13)$$
$$f(0, 0, \alpha) = 0, \qquad f(1, 0, \alpha) = 1, \qquad f(2, 0, \alpha) = 1, \qquad (14)$$
$$f(\alpha, 2, \beta) = 1. \qquad (15)$$

The function $F \colon \mathcal{A}^{\mathbb{Z}} \to \mathcal{A}^{\mathbb{Z}}$ defined by this CA is obviously a Lipschitz function. We prove that (12) holds when $d(x, y)$ is given by the Tychonoff distance (1). For $n > 0$ we define $x_n, \epsilon_n$ as follows. We set $m = \lceil n/2 \rceil$, $\epsilon_n = 2^{-m-1}$ and

$$x_n(i) = \begin{cases} 0, & \text{if } i > -m, \\ 1, & \text{if } i \leq -m. \end{cases}$$

Given $y$ such that $d(x_n, y) \leq \epsilon_n$, let $t_1, t_2$ be such that

$$x_n(i) = y(i) \ -t_1 < i < t_2, \qquad x_n(t_2) \neq y(t_2), \qquad x_n(-t_1) \neq y(-t_1).$$

In the following we assume that both $t_1$ and $t_2$ are finite but the same reasonig holds also if one of them is not finite. Note that our choice of $\epsilon_n$ implies that both $t_1$ and $t_2$ are greater than $m$. The foundamental observation is that the values $y(i)$ with $i \geq t_2$ or $i \leq -t_1$ do not affect the values $\big[F^k(y)\big](i)$ for $k \leq n$ and $-t_1 < i < t_2$ (see Fig. 3). Hence, for $k \leq n$

$$\begin{aligned} d\big(F^k(x), F^k(y)\big) &\leq \sum_{i \geq t_2} \frac{1}{2^i} + \sum_{i \geq t_1} \frac{1}{2^i} \\ &= 2\left(\frac{1}{2^{t_2}} + \frac{1}{2^{t_1}}\right) \\ &\leq 2d(x_n, y) \end{aligned}$$

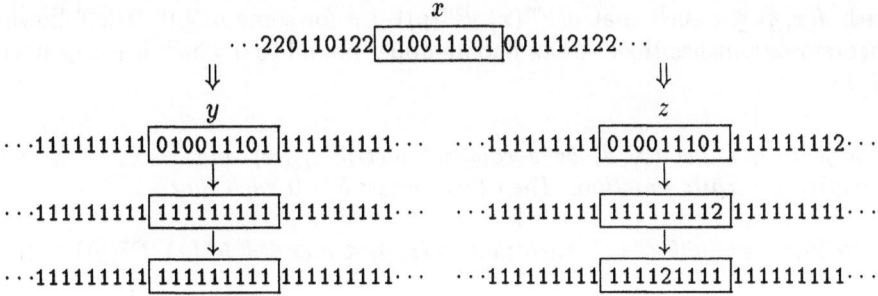

Figure 4: Sample configurations $x$, $y$ and $z$ for $m = 4$. We show $y$, $F^9(y)$, $F^{13}(y)$ (left) and $z$, $F^9(z)$, $F^{13}(z)$ (right).

which proves (12).

To prove that $F$ is sensitive (with constant $1/2$), we show that $\forall x \in \mathcal{A}^{\mathbf{Z}}$ and $\forall \epsilon$ there exist $y, z$ and an integer $n$ such that

$$d(x, y) \leq \epsilon, \quad d(x, z) \leq \epsilon, \quad d(F^n(y), F^n(z)) \geq 1. \qquad (16)$$

Given $0 < \epsilon < 1/2$, we choose $m$ such that $x(i) = y(i)$ for $|i| \leq m$ implies $d(x, y) < \epsilon$ (take for example $m = \lceil -4 \log_2 \epsilon \rceil$). We define $y$ and $z$ as follows (see also Fig. 4):

$$y(i) = \begin{cases} x(i), & \text{if } |i| \leq m \\ 1, & \text{if } |i| > m \end{cases} \qquad z(i) = \begin{cases} y(i), & \text{if } i \neq 3m + 1 \\ 2, & \text{if } i = 3m + 1 \end{cases}$$

To prove (16) we show that after $3m + 1$ steps we have $[F^{3m+1}(y)](0) = 1$ and $[F^{3m+1}(z)](0) = 2$. We first consider the simpler case in which $x(i) \neq 2$ for $|i| \leq m$ (see Fig. 4). After $2m + 1$ steps we have $[F^{2m+1}(y)](i) = 1$ for all $i$. In fact, since $y$ contains no 2's, by (13) $y(i) = 1 \implies [F^k(y)](i) = 1$ for all $k > 0$. Viceversa, since $f(1, 0, \alpha) = 1$, the number of 0's decreases at each step until none is left. Similarly, we have $[F^{2m+1}(z)](i) = 1$ for $i \neq m$, and $[F^{2m+1}(z)](m) = 2$ (since $f(\alpha, 1, 2) = 2$, the value 2 initially in position $3m + 1$ moves left by one position at each step). After $m$ more steps, we have $[F^{3m+1}(y)](0) = 1$, and $[F^{3m+1}(z)](0) = 2$ as claimed.

Now consider a generic $x \in \mathcal{A}^{\mathbf{Z}}$. By (14) we have that the number of 0's decreases by at least one at each step until none is left. Since $(\alpha, 1, 2)$ is the only triplet that generates a 2, we have that a symbol 2 "survives" only by moving left. Hence after $2m + 1$ steps we have $[F^{2m+i}(y)](i) = 1$ for $i \geq -m$. Similarly, $[F^{2m+1}(z)](i) = 1$ for $i \geq -m$ $i \neq m$, and $[F^{2m+1}(z)](m) = 2$. After $m$ more steps, we have $[F^{3m+1}(y)](0) = 1$ and $[F^{3m+1}(z)](0) = 2$ which proves (16). $\qquad \square$

The above example shows that the number of iterations required to double the distance between $x$ and an arbitrarily close point $y$ *does* depend on $x$. Note however, that by defintion of sensitivity we know that for all $\epsilon$ there exists $y$

with $d(x,y) \leq \epsilon$ such that $d(F^n(x), F^n(y)) > \delta$ for some $n \geq 0$. The following theorem establishes that we can find an upper bound to $n$ which is independent of $x$.

**Theorem 6.1** *Let $(X,d)$ be a compact metric space, and $F: X \to X$ be a sensitive Lipschitz function. Then there exists $\delta > 0$ such that*

$$\forall \epsilon \exists n_\epsilon: \quad \forall x \exists y \exists k \leq n_\epsilon \quad such \ that \quad d(x,y) \leq \epsilon \ and \ d(F^k(x), F^k(y)) > \delta.$$

**Proof.** Assume $F$ is a sensitive Lipschitz function with parameters $\delta'$ and $\lambda$. We prove the theorem for $\delta = \delta'/2$. Assume by contradiction that there exists $\epsilon$ such that

$$\forall n \exists x_n: \quad d(x_n, y) \leq \epsilon \quad \Longrightarrow \quad d(F^k(x_n), F^k(y)) \leq \delta \quad for \quad k \leq n. \quad (17)$$

Since $X$ is a compact space, we can find a sequence $x_n$ such that

$$\lim_{n \to \infty} x_n = \tilde{x}, \qquad d(x_n, \tilde{x}) \leq \delta \lambda^{-n}.$$

Since $F$ is sensitive with parameter $\delta' = 2\delta$, there exist $\tilde{y}$ and $m$ such that

$$d(\tilde{x}, \tilde{y}) \leq \epsilon/2 \quad and \quad d(F^m(\tilde{x}), F^m(\tilde{y})) > 2\delta$$

Let $i$ be the smallest integer such that $i \geq m$ and $d(x_i, \tilde{x}) \leq \epsilon/2$. We have

$$d(x_i, \tilde{y}) \leq d(x_i, \tilde{x}) + d(\tilde{x}, \tilde{y}) \leq \epsilon.$$

Moreover, by the triangle inequality we have

$$\begin{aligned}
d(F^m(\tilde{y}), F^m(x_i)) \quad &\geq \quad d(F^m(\tilde{y}), F^m(\tilde{x})) - d(F^m(\tilde{x}), F^m(x_i)), \\
&> \quad 2\delta - \lambda^m d(\tilde{x}, x_i) \\
&\geq \quad \delta.
\end{aligned}$$

Hence, the pair $x_i, \tilde{y}$ contradicts (17) and the theorem is proven. $\square$

# 7  Conclusions

In this paper we have considered two properties of DTDS, namely expansivity and positiveness of Lyapunov exponents, which are commonly used to detect chaotic behaviors. We have shown that for CA expansivity implies that Lyapunov exponents are positive and uniformly bounded away from zero. A surprising related result is that there are no expansive CA in dimension greater than one. Our proofs are based on properties of expansive Lipschitz functions over compact metric spaces which are of independent interest.

Our results show that expansivity is indeed a central notion for the characterization of chaotic behavior and that it deserves further investigation.

# References

[1] D. Assaf, IV and W. A. Coppel, Definition of Chaos. *Amer. Math. Monthly, 865*, 1992.

[2] J. Banks, J. Brooks, G. Cairns, G. Davis, and P. Stacey, On the Devaney's Definition of Chaos. *Amer. Math. Monthly, 332-334*, 1992.

[3] B. Codenotti and L. Margara, Transitive Cellular Automata are Sensitive. *Amer. Math. Monthly 58-62*, 1996.

[4] B. Codenotti and L. Margara, Chaos in Mathematics, Physics, and Computer Science: Similarities and dissimilarities. *Proc. of The Evolution of Complexity, Bruxelles*, May 1995.

[5] M. Martelli, Discrete Dynamical Systems and Chaos. *Pitman Monographs and Surveys in Pure and Applied Science 62, Longman Scientific & Technical*, 1994.

[6] R. L. Devaney, An Introduction to Chaotic Dynamical Systems. *Addison Wesley*, 1989.

[7] F. Fagnani and L. Margara, Expansive CA and Chaos. *Proc. of the fifth ICTCS*, Ravello, November 1995.

[8] P. Favati, G. Lotti, and L. Margara, Additive One Dimensional Cellular Automata are Chaotic According to Devaney's Definition of Chaos. *To appear in Theoretical Computer Science.*

[9] C. Knudsen, Aspects on Noninvertible Dynamics and Chaos. *Ph.D. Thesis*, 1994.

[10] C. Knudsen, Chaos Without Nonperiodicity. *Amer. Math. Monthly, 563-565*, 1994.

[11] L. Margara, Cellular Automata and Chaos. *Ph.D. Thesis*, 1995.

[12] M. Shirvani and T. D. Rogers, On Ergodic One-Dimensional Cellular Automata, *Commun. in Math. Phys. 136, 599-605*, 1991.

[13] M. A. Shereshevsky, Lyapunov exponents for one-dimensional cellular Automata, *J. Nonlinear Sci. 2, 1-8*, 1992.

[14] M. Vellekoop and R. Berglund, On Intervals, Transitivity = Chaos. *Amer. Math. Monthly, 353-355*, 1994.

# Quasiperiod-3 and period-3 collective behavior in three dimensional totalistic illegal cellular automata with high connectivity

F. Jiménez-Morales, H.Karma, J. J. Luque and M. C. Lemos

Departamento de Física de la Materia Condensada.

Universidad de Sevilla. P. O. Box 1065

41080-Sevilla, Spain

### Abstract

The collective behavior of totalistic illegal cellular automata rules in d=3 with high connectivity is studied. P3 and QP3 behavior of a family of rules is located obtaining the Shannon information function and using a mean-field approximation.For the QP3 behavior the time autocorrelation function decays as a power law with an exponent of -1/2.

PACS: 05.45-Theory and models of chaotic systems.

A cellular automata (CA) is a regular array of cells, each of which can take a finite number of values and evolves synchronously. The state of a cell at time t+1 is determined by the state of the cells in a neighborhood at time t and a local rule. CA are discrete dynamical systems of simple construction but complex behavior. Physical, chemical and biological systems with many discrete elements with local interactions can be modelled using CA[1].

An important question about CA is whether such systems can display periodic or quasiperiodic temporal oscillations (period-3 and quasiperiod-3) and under what circumstances. This behavior is neither transient nor due to the finite size of the lattice and has been obtained for deterministic and probabilistic rules. Recently it has been argued that this collective motion may be sustained through a fluctuating phase field that can be described by the Kardar-Parisi-Zhang (KPZ) equation[2, 3], and that there is a critical space dimension $d_c = 2$ above which globally coherent oscillations can be observed[4].

From work on finite-dimensional lattices[5] it was conjectured that the possibility of finding such behavior would increase the higher space dimension of the system and some cellular automata rules in d=4, d=5 and d=6 exhibit a period-3 and a quasiperiod-3. In d=3, where it is easier to simulate a real physical system, rule-33 is the only one with such behavior[6].

For the appearance of non-trivial collective behavior the topology of the connections plays an important role as well as connectivity and space dimension [4, 5, 7]. In previous papers, where non-trivial collective behavior was found, the neighborhood selected is mainly the Von Neumann neighborhood that consists of nearest neighbours. But it has been found that increasing the connectivity new rules show the quasiperiod and period-3 behavior in ordered lattices[8] , and in random connected cellular automata there is a critical connectivity $(N = 13)$ at which the number of rules with non-trivial behavior has a maximum value [9]. In this paper we search for new rules with non-trivial collective behavior (period-3 and quasiperiod-3) in d=3 with a neighborhood that consists of the cell $(i, j, k)$ plus the cells $(i\pm1, j, k)$, $(i\pm2, j, k)$, $(i, j\pm1, k)$, $(i, j \pm 2, k)$, $(i, j, k \pm 1)$, $(i, j, k \pm 2)$ .

Considering a $d = 3$ lattice, the following definitions will be useful:

- $X_{ijk}(t)$ is the state (0 or 1 ) of the cell with indices $(i, j, k)$ at time t.

- $S_{ijk}(t)$ is the sum of the state of the neighbors plus the cell $(i, j, k)$ at time t. The considered local rules are totalistic , i.e., the state of a cell in time $t + 1$ depends only on the value of $S_{ijk}(t)$ . Also only illegal rules will be studied because the only cellular automata rule in three dimensions which was known to show a quasiperiod three behavior was rule 33 $(2^5 + 2^0 = 33$ ). To simplify the notation each rule will be denoted by $R_{\{S\}}$ where $\{S\}$ is a set of integer values $S \in [1, N - 1]$ , and $N$ is the connectivity. The rules are defined by:

$$X_{ijk}(t+1) = \begin{cases} 1 & if \ S_{ijk}(t) = 0 \ or \ S_{ijk}(t) \in \{S\} \\ 0 & otherwise \end{cases}$$

The number of possible rules is $2^{N-1}$ but taking care of the symmetries this number is reduced to $2^{N-2}$ and as some rules are autosymmetric the number of studied rules is 1979 . Periodic boundary conditions will be used.

- The global behavior of the CA will be monitor through the concentration of activated cells at time t , $c(t) = \frac{1}{n} \sum_{i,j,k=1}^{n} X_{ijk}(t)$ , where n is the number of cells.

A reticular lattice of $60^3$ cells is taken. The system evolves during 5000 time steps and the concentration is measured in the last 1000 time steps. When a rule shows the QP3 or P3 the lattice size is increased up to $100^3$ cells and the system evolves during $10^4$ time steps. The initial concentration is selected randomly with a value of 0.5 .

According to the time series of the concentration the types of collective behavior found are: noisy period-1, period-2, quasiperiod-2, period-3 (P3) and quasiperiod three (QP3). Rules that exhibit P3 and QP3 behavior are chaotic in the sense of Wolfram´s class III. Each site of the lattice follows a chaotic evolution which has no apparent relation to the global one. The global variables do show fluctuations that decrease as the lattice size increases leading to a well defined thermodynamic limit. The results are summarized in Table I for the different families of rules that show P3 and QP3. In Figure-1 a typical QP3 and P3 behavior are shown. The phenomenology of these behaviors has already been extensively studied[10].

| | P3 | QP3 |
|---|---|---|
| $R_{\{S1\}}$ | | $\{6\},\{7\}$ |
| $R_{\{S1-S2\}}$ | $\{6,9\text{-}11\},\{7,9\text{-}11\},\{8,10\text{-}11\},$ $\{9,10\text{-}11\},\{10,11\}$ | $\{6,8\},\{6,12\},$ $\{7,8\},\{7,12\},\{8,9\}$ |
| $R_{\{S1-S2-S3\}}$ | $\{6,9,11\text{-}12\},\{6,10,11\text{-}12\},\{6,11,12\},$ $\{7,9,12\},\{7,10,11\text{-}12\},\{7,11,12\},$ $\{8,9,12\},\{8,10,11\text{-}12\},\{8,11,12\},$ $\{9,10,11\text{-}12\},\{10,11,12\}$ | $\{6,8,12\},\{6,9,10\},$ $\{7,8,10\text{-}12\},\{7,9,10\text{-}11\},\{8,9,10\text{-}11\}$ |
| $R_{\{S1-S2-S3-S4\}}$ | $\{6,9,11,12\},\{7,10,11,12\},$ $\{8,10,11,12\},\{9,10,11,12\}$ | $\{1,7,8,11\text{-}12\},\{1,7,9,11\text{-}12\},\{1,7,10,11\text{-}12\},$ $\{1,8,9,10\text{-}11\},\{1,8,9,12\},\{1,8,10,11\text{-}12\},$ $\{1,9,10,11\text{-}12\},\{6,9,10,12\},\{7,8,11,12\},$ $\{7,9,10,12\},\{7,9,11,12\},\{8,9,10,12\},\{8,9,11,12\}$ |
| $R_{\{S1-S2-S3-S4-S5\}}$ | | $\{1,2,8,9,10\text{-}12\},\{1,2,8,10,11\text{-}12\},\{1,2,9,10,11\text{-}12\},$ $\{1,2,10,11\text{-}12\},\{1,5,6,7,8\},\{1,7,9,11,12\},\{1,7,10,11,12\},$ $\{1,8,10,11,12\},\{1,9,10,11,12\},\{4,5,6,7,8\}$ |
| $R_{\{S1-S2-S3-S4-S5-S6\}}$ | | $\{1,2,3,8,9,11\},\{1,2,3,8,10,11\},$ $\{1,2,3,9,10,11\text{-}12\},\{1,2,3,10,11,12\}$ |

Table I: Illegal Totalistic Rules in d=3 with a connectivity of N=13 that show non-trivial collective behavior P3 and QP3. Lattice size is $100^3$ for rules $R_{\{S1\}}$ and $R_{\{S1-S2\}}$ and $60^3$ for the other ones.

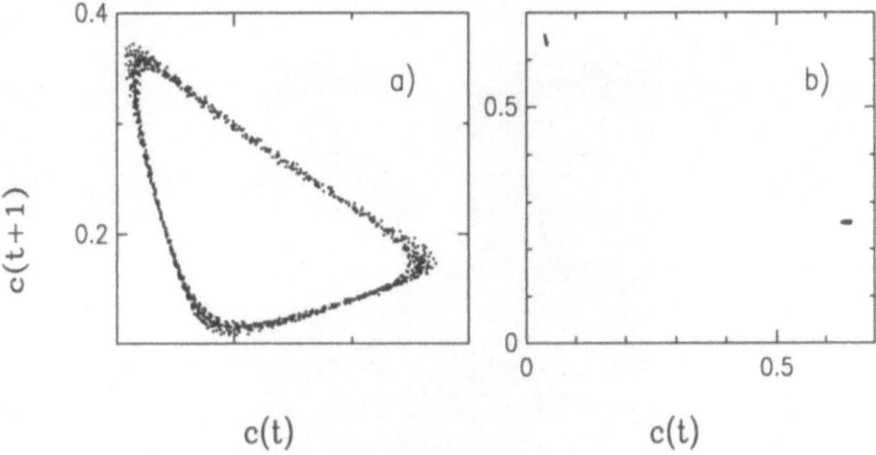

Figure 1: The iterative map of new rules that show a QP3 and a P3 in d=3 with a connectivity of 13 cells, starting from a random initial concentration of 0.5. Lattice size is $100^3$. Transient discarded. (a) Rule $R_{\{1-2-9-10-11\}}$. (b) Rule $R_{\{8-10-12\}}$.

Figure 2 shows a contour plot of the different behavior found in the real simulations ( from white to dark: QP3,P3,QP2,P2 and P1) for the family of rules $R_{\{S1\}}$ and $R_{\{S1-S2\}}$. It can be seen that most of the rules with P3 and QP3 are locate in the region $S1 = 6$, $S2 \in [6, 12]$ .

An approach to locate the region of most interesting behavior can be done measuring the Shannon information function defined as:

$$H = -K \sum_{i=1}^{n} p_i \log p_i. \tag{1}$$

Where $p_i$ is the probability of occurrence of state "i". Within a mean-field approximation we can obtain the probability that a cell will be 1 at time t+1 as follows:

$$y = (1 - x)^N + \binom{N}{S1} x^{S1}(1 - x)^{(N-S1)} + a\binom{N}{S2} x^{S2}(1 - x)^{(N-S2)} \tag{2}$$

where "y" is the concentration at time t+1 and "x" is the concentration at time t, $a = 0$ for the family of rules $R_{\{S1\}}$ and $a = 1$ for $R_{\{S1-S2\}}$ . The procedure to obtain $H$ is as follows:

a) We calculate the intersection point ($x_{int}$ ) between the iterative map and the line $y = x$; b) the equation (2) is iterated and the dynamics is reduced to a binary sequence 0 if $x < x_{int}$ and 1 if $x \geq x_{int}$; c) a string of 8 digits is considered a state thus we have $2^8$ states; d) After a transient of 1000 steps

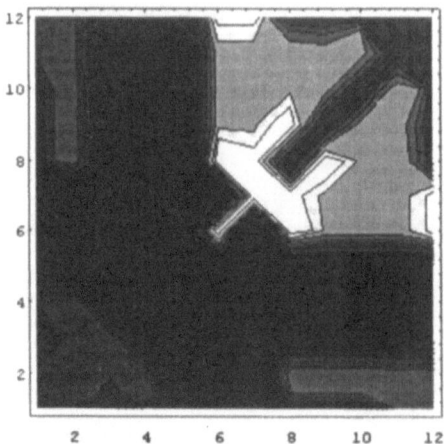

Figure 2: Contour Plot of the different behaviors obtained in the CA simulations for the family of rules $R_{\{S1\}}$ and $R_{\{S1-S2\}}$. White corresponds to QP3 and black to P1.

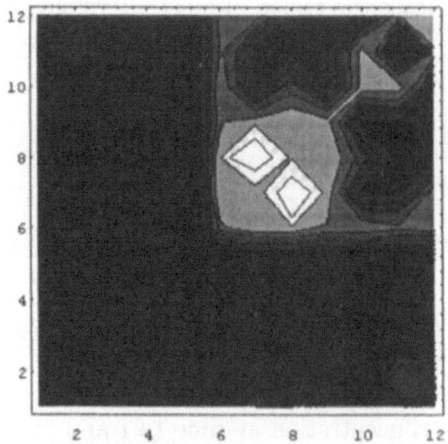

Figure 3: Contour Plot of the Shannon information function H for the family of rules $R_{\{S1\}}$ and $R_{\{S1-S2\}}$. White is the highest value (QP3) and black (P1) the lowest one.

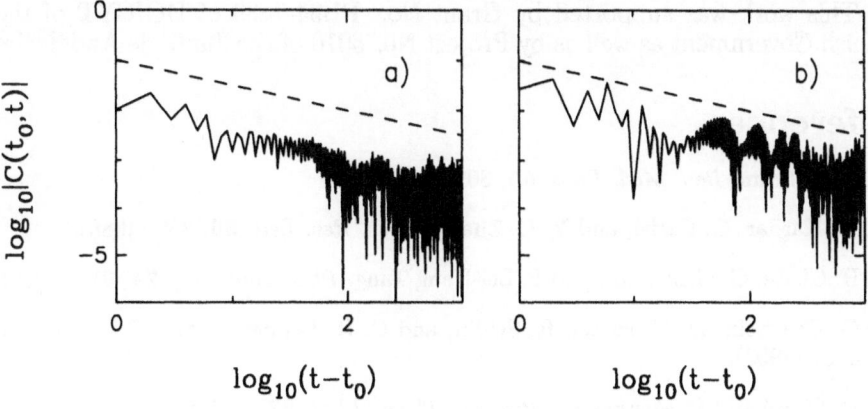

Figure 4: Log-log plot of the absolute value of the time autocorrelation function. (a) Rule $R_{\{6\}}$. (b) Rule $R_{\{7\}}$. The slope of the dashed lines is -1/2.

the probability $p_i$ of each state is measured when the system is running over $2^{16}$ time steps.

Figure 3 shows the contour plot of $H$. The maximum values appear in the region of $S1 = 6$ and $S2 \in [6, 12]$ where the real CA shows the P3 and QP3 behavior.

Finally the time autocorrelation function was studied, defined by

$$C(t_0, t) = \frac{1}{n} \sum_{ijk}^{n} [X_{ijk}(t_0).X_{ijk}(t)] - c(t_0)c(t)$$

The absolute value of $C(t_0, t)$ is shown in Figure 4 for rules $R_{\{6\}}$ and $R_{\{7\}}$. $C(t_0, t)$ oscillates in t and the envelope of the oscillation decays as $(t-t_0)^{-(d-2)/2}$. This provides more numerical support for the hypothesis that phase fluctuations in these systems are described by the KPZ equation [3].

In summary, we have done numerical simulations in d=3 of a family of illegal cellular automata with a high connectivity and found new rules that show the striking collective behavior QP3 and P3. Therefore we extend the number of models that can be used for the simulation of real systems with this behavior. In the space of the rules the most interesting ones can be located using the Shannon information function. Finally for the QP3 behavior we provide more numerical evidence that the autospin time-correlation function decays with time as predicted by the Kardar-Parisi-Zhang equation.

This work was supported by Grant No. PB94-1439 of DGICYT of the Spanish Government as well as by Project No. 6010 of the Junta de Andalucia.

# References

[1] S. Wolfram. *Rev. Mod. Phys.* **55**, 601 (1983).

[2] M. Kardar, G. Parisi, and Y. C. Zhang. *Phys. Rev. Lett.* **56**, 889 (1986).

[3] H. Chaté, G. Grinstein , and P. Lei-Hang Tang. *Phys. Rev. Lett.* **74**, 912 (1995).

[4] G. Grinstein, D. Mukamel, R. Seidin, and C. H. Bennett. *Phys. Rev. Lett.* **70**, 3607 (1993).

[5] H. Chaté and P. Manneville. *Progress Theor. Phys.* **87**, 1 (1992).

[6] J. Hemmingsson. *Physica A* **183**, 255 (1992).

[7] F. Jiménez-Morales and J. J. Luque. *Physica A* **212**, 118 (1994).

[8] F. Jiménez-Morales et al. (in preparation).

[9] N. Mousseau. *Europhys. Lett.* **33**, 509 (1996).

[10] H.Chaté, A.Lemaître, Ph.Marcq, P.Manneville. *Physica A* **224** (1996).

# Simulation of water flow through a porous soil by a Cellular Automaton model

S. Di Gregorio°, R. Rongo°, R. Serra*, W. Spataro°, M. Villani*
*CRA Montecatini via Menotti, 48 48023 Marina di Ravenna (RA), Italy
e-mail rserra.cramont@arcobaleno.com
°Dept. of Mathematics, Univ. of Calabria, 87036 Arcavacata di Rende (CS), Italy
e-mail toti.dig@unical.it

### Abstract

Soil bioremediation is a highly complex phenomenon and involves several disciplines at the same time, including fluid dynamics, chemistry and biology. A cellular automata model is currently under development, which deals with all these kinds of phenomena. In this paper the fluid dynamical aspects of this model are discussed, and some comparisons with experimental data are presented. Genetic algorithms have been applied in order to tune the model to a specific case.

## 1 Introduction

Although less debated than air and water pollution, soil pollution, either due to industrial or agricultural activities, is one of the major environmental problems in industrial countries. Biological methods are at the root of some of the best techniques to recover contaminated land [Baker, 1994]. In particular, in-situ bioremediation is based upon the use of bacteria to degrade the contaminant and transform it into less dangerous compounds (ultimately, this often leads to water and carbon dioxide).
Since it involves several physical, chemical and biological phenomena, bioremediation is a complex process, and several laboratory and pilot plant experiments are required, before starting the field operations.
A first important goal in order to face soil pollution problems is the ability to simulate water flow through the soil. This paper concerns our investigation about water infiltration in a porous soil; the cellular automata model, which is compared with the results of careful experiments performed in the laboratories of CRA Montecatini, is a part of a wider project which deals also with the chemical and biological phenomena. Model development is achieved by a continuous interaction loop between simulations, design and execution of new experiments and model improvements.

## 2 Microscopic and macroscopic Cellular Automata

Modelling complex physical systems is today a major challenge of theoretical research, which has at the same time important practical applications. Actually, most "interesting" systems belong to this class, and the limitations of traditional approaches have been pointed out by a number of authors [Haken, 1978; Prigogine, 1969; Serra et al., 1986]. Cellular automata (shortly, CA) have been very useful for exploring the features of a wide class of complex systems. In particular, their application to physical systems has brought important results [Toffoli 1987; WoObram 1986].
Most physical applications of CA's rely upon the use of a "microscopic" approach, where discrete particles are the basic tokens, and the familiar laws of continuum mechanics, like e.g. the continuity equation, or the Navier-Stokes equations, are obtained with some appropriate limiting operations

[Toffoli, 1994]. Physical phenomena concerning our problem have been treated successfully by formal support of CA [Chopard & Droz 1991, Karapiperis & Blankleider 1993], but they are considered at a microscopic level, which can involve severe problems, when applications concern a macroscopic scale.

A different approach is also possible to modelling physical systems with CA's, where the state space of a single cell is allowed to be the Cartesian product of a number of spaces, and where variables can take either a large set of possible values, or even continuous values. Such "mesoscopic" or "macroscopic" CA's (as opposed to the classical, microscopic ones) have been successfully applied to a number of physical phenomena, including the simulation of lava flows [Barca et al., 1994] and rock fracturing [Norman et al., 1991].

Here we propose a not microscopic CA approach, where each cell is characterised by specific values (the state) of selected physical variables, which can be of a macroscopic kind, like e.g. water content or contaminant concentration.

## 3  The layered structure of the model

The experiment consists in flooding a container having dimensions 40x125x50 cm., filled with 350 Kg. of soil with a phenol aqueous solution, and in the subsequent phenol degradation performed by the indigenous bacteria. The most relevant physical variables for the contamination phase are the water and phenol distribution inside the container and the drainage water collected at the bottom of the container. The flooding solution was 15 lt. of water with 212 g. of phenol.

The simulation of bioremediation involves therefore a number of phenomena, which are physical (e.g. multiphase flow in a porous medium), chemical (contaminant degradation, interaction with chemicals on the pore walls, etc.) and biological (like biomass growth, interaction among bacterial populations, etc.) in nature.

The CA model reflects this distinction by presenting 3 layers: a fluid dynamical layer, which describes the fluid flow through the medium, a "chemical" layer, which describes the transport of pollutant and/or nutrients and their chemical reactions, and a biological layer, which describes bacterial dynamics and their interaction with water, pollutants and nutrients.

The cell size is large enough to include a large number of pores, so that the fluid dynamical variables of each cell represent averages taken over of a large number of pores. The choice of such an intermediate description level allows the model to describe significant spatial variations and dishomogeneities while avoiding the burden of microscopic details.

The purpose of the contamination model (i.e. the first two layers) is the simulation of the behaviour of a vessel filled with soil, flooded with an aqueous phenol solution. The third layer simulates the subsequent remediation, which is achieved by spraying the soil with a nutrient rich solution. The time behaviour of several variables have been measured experimentally [Banti et al, 1996], including the percolated water flow and the contaminant concentrations in drainage water and in soil.

## 4  Fluid flow through a porous medium

Our attention will be directed to the main aspects of the behaviour of the mobile water inside the soil, which is considered as a solid matrix, rigid and stable, without consolidation or subsidence.    Water saturation (volume of water / volume of voids) is crucial in order to approach the problem; at the lowest level of saturation a minimum quantity of water, the irreducible water, surrounds the grains of soil, it is immobile and can be removed only in the evaporation process by strong heating of the soil. At slowly greater saturation levels, water forms rings (pendular rings) around the grain contact points without forming a continuous water phase, except a thin film, influenced essentially by intermolecular forces.

As water saturation increases, the pendular rings expand until a water continuum is reached (water funicular saturation), so that water can flow; if water saturation is further increased, all the pores will be filled, except air, if any, entrapped in the largest pores.

Let's consider water flow from a phenomenological viewpoint. We are interested in the process from the beginning of the water funicular saturation (equilibrium water saturation) to conditions of saturation. In a first phase we can omit any description of water forming thin films around the grains, considering it irreducible.

As mobile water enters a soil, flow paths are created; at the beginning, with lowest saturation, the paths are very tortuous; an obstacle to the water flow is given by the interactions at the solid-water interface, generating high viscosity also by the capillary forces, which produce suction/retention effects.

As the saturation increases, the flow takes place through larger pores; this process causes both a sensible increase in the cross sectional area and a reduction in the path tortuosity, reducing drastically the effects of the interactions at the solid-water interface and thus lowering the effective viscosity.

The parameter, which describes the complex behaviour resulting from the different factors in the continuity equations, is the hydraulic conductivity, whose values range from approximately 0 with only irreducible water content to a maximal value $K_0$ at the full saturation point (Freze and Cherry, 1979; Bear 1979).

This process can reach its limit, when full saturation occurs; the inverse process can be described in an analogous way, considering that in the draining phase from full saturation to equilibrium water saturation a hysteresis effect is evidenced for the hydraulic conductivity, whose value is larger in wetting than in drying.

If isothermal conditions are given and the hysteresis effects are negligible, then an analytical expression, given by Buckingam (1907) and following Darcy's law, can be adopted as a reference in the development of the model [Mendicino 1993]

$$\overline{q} = -K(w)\overline{\nabla}\phi$$

where $\overline{q}$ is the flow rate, $w$ is the water content and $K$ is the hydraulic conductivity, $\Phi = z + \Psi(w)$ represents the hydraulic head, with $z$ the elevation and $y$ the pressure head.

The water transport outside a cell will be given by two components, the former one, the diffusion flow, which tries to reduce the difference between $\psi$ of the central cell and that of its neighbours, the latter one determined by the effect of gravity force and the value of the hydraulic conductivity.

# 5 SOIL generalities

The CA specification is given by:

$$SOIL = (R_3, A_1, A_{6a}, A_{6b}, X, Q, \sigma, \gamma_1, \gamma_{6a}, \gamma_{6b})$$

where

- $R_3 = \{(x, y, z) | x, y, z \in N, 0 \le x \le l_x, 0 \le y \le l_y, 0 \le z \le l_z\}$ is the set of points with integer co-ordinates in the finite region, where the phenomenon evolves. $N$ is the set of natural numbers.
- $A_1 \subset R_3$, specifies the water source cells at air-soil contact, where contamination begins; such cells obey to the particular transition function $g_1$ substitutive, but compatible with $\sigma$. In cells of $A_1$ the contaminated water appears.
- $A_{6a}$, $A_{6b} \subset R_3$, specify the two last down layers of cells, which obey respectively the particular transition functions $\gamma_{6a}$, $\gamma_{6b}$ different from $\sigma$.
- The set X identifies the geometrical pattern of cells which influence the cell state change. They are the cell itself and the "up", "north", "east", "west", "south" and "down" neighbouring cells, which are individuated respectively by the indexes 0, 1, 2, 3, 4, 5, 6:

$$X = \{(0, 0, 0), (0, 0, 1), (0, 1, 0), (1, 0, 0), (-1, 0, 0), (0, -1, 0), (0, 0, -1)\};$$

The finite set Q of states of the elementary automaton is:

$$Q = Q_w \times Q_p \times Q_{outf6} \times Q_{K\_sat} \times Q_{cap\_thr} \times Q_{cap\_rate} \times Q_{grav\_rate} \times Q_{cap\_power} \times Q_{grav\_power}$$

whose components represent the substates:

$Q_w$ is the water content in the cell.

$Q_p$ is the "effective porosity parameter". It depends on physical-chemical characteristics of the soil in the cell and individuates the quantity of water drainable from the cell in condition of full saturation, i.e. the porous volume, which can be filled by water.

$Q_{outf}$ represents the water flow toward the six neighbourhood directions from the central cell. Note that the inflows are not explicitly considered, they are obtained trivially by the outflows.

$Q_{K\_sat}$ is the hydraulic conductivity at the saturation, considering that, inside the cell, the soil is homogeneous.

$Q_{cap\_thr}$ is the "capillary water threshold" substate, i.e. the maximum cell water, which is effected by the capillary forces.

$Q_{cap\_rate}$ and $Q_{grav\_rate}$ are parameters ruling the rate of the capillary and gravitational water outflows, their values are interdependent and bound by the clock of CA.

$Q_{cap\_power}$ and $Q_{grav\_power}$ define a multiplicative factor represented by the power of the partial cell saturation, which effect the rate of the capillary and gravitational water outflows.

- $\sigma : Q^7 \rightarrow Q$ is the deterministic state transition function for the cells in $R_3$, an outline of its specification will be given in the next sections.

- $\gamma_1 : Q^2 \times N \rightarrow Q$ is the transition function of source cells, where water has "origin" at each SOIL time interval $t \in N$; according to the assigned feeding, the outflow to the inferior cell only is calculated (recall that "1" is the up direction).

- $\gamma_{6a} : Q^2 \rightarrow Q$ is the deterministic state transition function for the cells in the first down layer, it specifies the behaviour of ideal cells, which receive water only from the superior cell and transmit it to the inferior one without retention (recall that "6" is the down direction).

- $\gamma_{6b} : Q^2 \rightarrow Q$ is the deterministic state transition function for the cells in the second down layer, it specifies the behaviour of ideal cells, which receive unlimited water without transmit it at all.

  Such ideal cells with strange transition functions $\gamma_{6a}$, $\gamma_{6b}$ are considered in order to describe forced conditions in laboratory experimental devices.

At the CA step 0 the initial configuration is defined, specifying all the starting values of the cell substates. At each next step, the function $\sigma$ is applied to all cells in $R_3 - \{A_1 \cup A_{6a} \cup A_{6b}\}$, while at the same time step the functions $\gamma_1$, $\gamma_{6a}$, $\gamma_{6b}$ are respectively applied to $A_1$, $A_{6a}$, $A_{6b}$, so the configuration is changed and an evolution step of SOIL is obtained.


# 6 The fluid flow through the soil


## 6.1 The capillary water flow

Capillary forces involve a suction effect, which covers the porous surface with a water layer, whose maximum possible thickness depends on the attractive inter-molecular forces between water, air and soil. We can characterise the equilibrium by a suction threshold or capillarity threshold ("cap_thr"), which is the minimum water quantity inside the cell, when the suction effects are practically negligible. Part of outflows ("new_outfc") from the central cell to its neighbours are originated because of balancing the capillary forces, when "cap_thr" is not reached in the neighbours.

```
procedure new_capillary_water_flow;
........
for(i=1;i<=6;i++)
 if (w[i]>suct_thr[i]){
  eliminated[i]=true;
  suct_ind[i]=1-w[i]/suct_thr[i];
  }else
  eliminated[i]=false;
do{
 new_control=false;
 sum_w=w[0];
 sum_suct_thr=suct_thr[0];
 for(i=1;i<=6;i++)
  if(!eliminated[i]){
   sum_w=sum_w+w[i];
   sum_suct_thr=sum_suct_thr+suct_thr[i];}
 av_suct_ind=1-sum_w/sum_suct_thr;
 for(i=1;i<=6;i++)
  if(suct_ind[i]>av_suct_ind)AND(!eliminated[i]){
   new_control=true;
   eliminated[i]=true;}
 }while(!new_control);
```

The capillary forces can be considered approximately proportional to the suction index "suct_ind", here defined as (cap_thr-w)/cap_thr, when  w < cap_thr  (0  otherwise); so a balance can be given by outflows, which minimise the total difference of the suction index between a cell (the central cell) and its six neighbours after each time interval.

In order to describe the long tail of the data about drainage water, it is supposed that the speed of the flow of capillary water depends upon the saturation (i.e. water content) of the cell. This hypothesis has a sound physical basis: let it suffice to recall that it concerns a thin layer of water on the pore surface, and that such flow should be slower at very low saturation levels (when not all the surface is already wet by water) and faster at high saturation levels. Analogous dependencies of diffusion upon saturation are found in most PDE macroscopic models of flow through porous media, where they take the form of a saturation dependent diffusion coefficient.  This dependence has been introduced by using an S-shaped function of the ratio water held in the cell/ total porosity (function "gf()"); this function is approximated by a broken line composed by three straight lines.

Computation involves four sequential actions:

1) Individuation of the neighbouring cells (the so-called "eliminated cells"), toward which capillary water outflow is not possible, because their suction index is larger than the suction index of the central cell or because other neighbouring cells have by comparison, such a low suction index that the capillary water flows necessarily only toward such cells.

2) Computing the values of the outflows toward the not eliminated cells, in order that the same value of the suction index is assumed by the central cell and the not eliminated cells.

3) Correction of the outflow computed values according to the time constant of the process; at each time step only a "cap_rate" fraction of the calculated flow is considered to pass from the central cell to the neighbour in order to take into account the kinetic constant of the capillary flow.

4) Correction of the outflow computed values according to function gf().

## 6.2 The gravitational water flow

The gravitational water in a cell is the part of water "gw", which is practically not sensible to capillary forces and is driven downwards by gravity. Its flow is ruled by the hydraulic conductivity, which depends upon the hydraulic conductivity at the saturation "K_sat", the porosity "p" and the water content "w" according to a function "f".
In order to calculate the flow between a cell and the cell below, the conductivity value, which is considered, is the minimum of the conductivities of the two cells, "K_min".

```
procedure new_vertical_gravitat_flow;
............
if (K_sat[0]<K_sat[6])
 K_sat_min=K_sat[0]
else
 K_sat_min=K_sat[6];
max_flow[6]=gw[0]*K_sat_min;
rec[6]=p[6]-w[6]-outf_sum[6];
if (max_flow[6]<rec[6])
 new_outfg[6]=rec[6]/grav_rate
else{
 new_outfg[6]=max_flow[6]/grav_rate
new_outfg[6]=new_outfg[6]*gf(w[0]/p[0])}
```

The above C-like procedure illustrates the algorithm for the computation of the new vertical gravitational flow "new_outfg"; "max_flow" is the maximum possible gravitational outflow from the cell, "p" is the cell porosity, "rec" is the maximum water quantity, which can penetrate a cell, partially filled. A first correction of the outflow computed values is made in order to account for the clock (time interval corresponding to a CA step) of SOIL; only a "grav_rate" fraction of the calculated flow is considered to pass from the central cell to the neighbour in order to take into account the kinetic constant of the capillary flow. A second correction of the outflow computed values must be made according to the same function of capillarity.

# 7. The search for optimal parameter: the Genetic Algorithm

The parameters that rule the behaviour of the proposed cellular automata may be divided in two categories: the physical ones - which are at least in principle directly measurable with a series of experiments - and the phenomenological ones, which are used in the model but which may lack a direct physical meaning. These latter parameters can be chosen in such a way as to adapt as much as possible the output of the algorithm to the experimental data, i.e. to minimise the distance between the predicted and the observed data. It can be mentioned that physical parameters include e.g. soil porosity and permeability, residual water content, etc., while adjustable parameters include e.g. a threshold which separates the water which is mainly subject to gravity from capillary water. Further adjustable parameters are those which describe the dependence of the water flow velocity from the cell's water content, and others.
The search for optimal parameter values can be done with optimisation algorithms. Genetic algorithms, which are a particular class of general problem solving algorithms inspired by biological evolution and

which are not strictly deterministic but involve stochasticity, have been chosen [see Holland 1975; Goldberg, 1989; Serra & Zanarini, 1990 for a quicker review].

There are many different forms of genetic algorithms, but the three genetic fundamental operators, whose use is widespread, are selection, crossover and mutation. The G.A. which has been used is slightly more sophisticated than the "bare" one; in particular, a two dimensional topology has been introduced, which mimics the spatial distribution of the population (see Figure 1). Competition and genetics act only among neighbouring individuals, while slow migration between different zones is also present. In this way "ecological niches" are created, and it is easier to avoid the danger that some relatively good individual spreads rapidly through the population ("premature convergence").

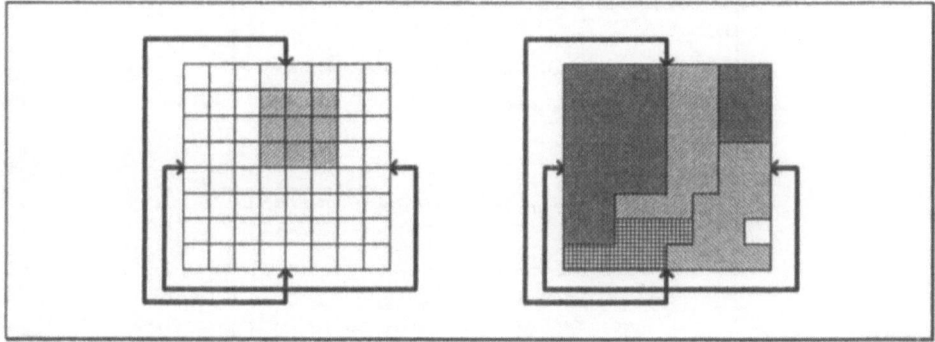

Figure 1. Topology of the population (left) and an example of genetically homogeneous niches (right)

## 8. 1-dimensional and 3-dimensional results

The model developed here is three dimensional: this kind of models is necessary in order to deal with real field operations, where spatial homogeneity is generally lacking. As the experimental apparatus is (macroscopically) homogeneous in the horizontal plane, most of the experimental data refer to the vertical dimension only. So it has been considered interesting to use also a 1-D version of the model which captures most of the interesting physics, and to compare it with the experimental data.

The 3-D model differs from the 1-D model mainly because of the presence of horizontal water diffusion. It will be shown that the 3-D model allows a direct introduction of preferential paths for water flow, which can be simulated in 1-D models only by introducing a further mechanism for water motion, i.e. through fissures. The fissures are macroscopic cavities which allow a faster fall of the cell water. Such mechanism has sound physical basis, and is known in the literature about agricultural soils. Fissures can be found in most soils; in our case they may have been partly originated by the initial pilot plant settlement and perhaps partly by the process of extracting some amounts of soil for chemical and biological analyses. In the model the fissural water is created only when the cell is almost saturated, and often this it begins a path which, due to dead ends, can go on only by diffusion or transport. We have created a preferential channel in which most part of the liquid is lost during the passage from a cell to another, which favours the other types of transport.

Some model results can be seen in Figure 2.

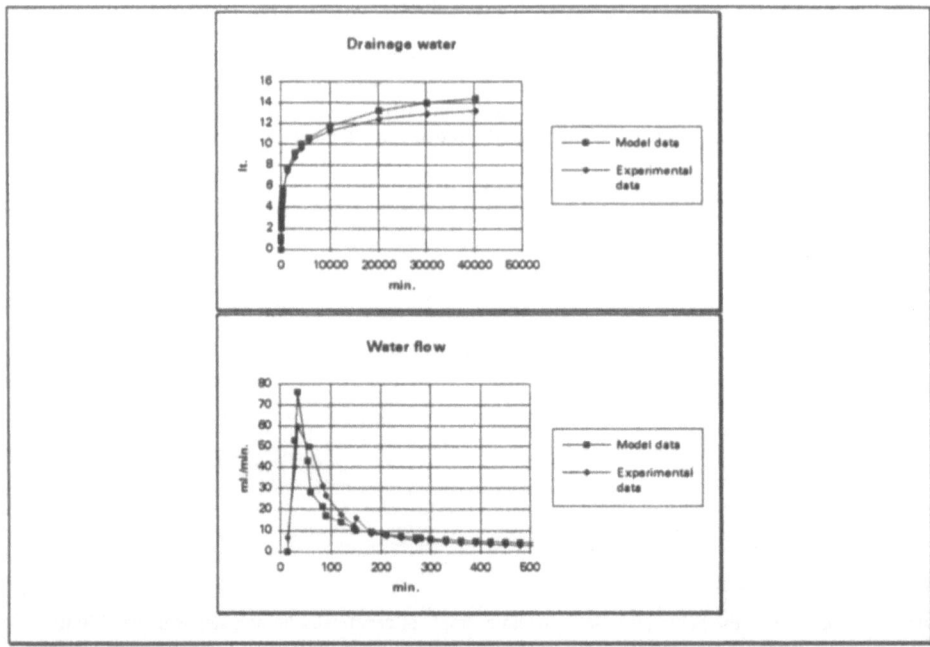

Figure 2. Comparison of model results and experimental data for water flow

In the 3-D case preferential paths for the fall of water through the soil can be simulated by a proper horizontal distribution of initial saturation, which mimicks the observed ponds which are formed in some experimental conditions. A vertical section of the 3-D automata, where the preferential paths formed by puddles and internal dishomogeneities could be observed, is shown in Figure 3.

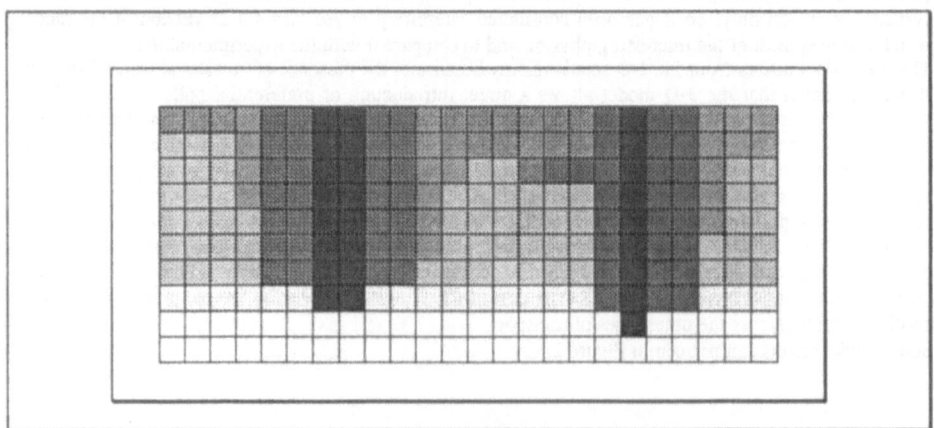

Figure 3. The preferential paths formed by puddles and internal dishomogeneities

# 9. Conclusions

The model presented here has proven able to successfully describe a variety of experimental data. Although it has several adjustable parameters, it is not trivial to describe phenomena which present widely different time scales, ranging from the few hours of the fast outflow to the very slow tail of percolating water. Moreover, the model can also reproduce the behaviour of different operating conditions, e.g. sudden flooding of the vessel vs. slow rain falling upon its upper layer.

The development of a reliable fluid dynamical model is a necessary prerequisite for a more complete bioremediation model (including chemical and biological phenomena).

The use of a macroscopic CA model, in this case, closely parallels the classical approach to fluid flow through a porous medium. The adoption of a CA framework can make it easier to describe in a unified way also the chemical and biological phenomena.

# Acknowledgements

This work has been partly funded by the UE Esprit-Capri initiative, subproject Caboto (9452 94 193 70). We gratefully acknowledge the experimental work performed by our colleagues at CRA, which allowed a severe testing of the model which led to major improvements, and the availability of the Camel software environment developed at CRAI.

# References

1. Baker KH. Herson D.S. Bioremediation. McGraw-Hill, New York, 1994

2. Haken H. Synergetics. An introduction: nonequilibrium phase transitions and self-organisation in physics, chemistry and biology. Springer-Verlag, Heidelberg, 1978

3. Prigogine I. Introduction to thermodynamics of irreversible processes. Wiley, 1969

4. Serra et al. Introduction to the physics of complex systems. Pergamon Press, Oxford, 1986

5. Toffoli T, Margolus N. Cellular automata machines. The MIT Press, Cambridge, Massachusetts, 1987

6. Wolfram S. Theory and application of cellular automata. World Scientific, Singapore, 1986

7. Toffoli T. Occam, Turing, von Neumann, Jaynes. How much can you get for how little? (A conceptual introduction to cellular automata). In Proceedings of the Conference ACRI'94, Rende (CS), 1994

8. Chopard B, Droz M. Cellular Automaton model for the Diffusion Equation. Journal of Statistical Physics 1991; 64 (3/4):859-892

9. Karapiperis T, Blankleider B. Cellular Automaton model of mass transport with chemical reactions. Bericht Nr.93-06 Paul Scherrer Institut, Villingen, Schweitz, 1993

10. Barca D et al. Cellular Automata for simulating lava flow: a method and examples of the etnean eruptions. Transport theory and statistical physics 1994; 23(1-3):195-232

11. Norman MG. The use of the CAPE environment in the simulation of rock fracturing. Concurrency: Practice and Experience, 1991; 3 (6):687-698

12. Banti et al. A detailed analysis of in-situ bioremediation: pilot plant experiments with a phenol contaminated soil. In Proceedings of the International Conference "Water Resources and the Environment", Cernobbio (CO), 1996

13. Freze RA, Cherry JA. Ground water. Prentice Hall, 1979

14. Bear J. Hydraulics of ground water. Mc Graw Hill, Inc., New York, 1979

15. Buckingam E. Studies on the Movement of Soil Moisture. U.S. Dept. Agr. Soil Bull.38, 1907

16. Mendicino G. Idrologia delle perdite. Pàtron Ed., Bologna, 1993

17. Holland JH. Adaptation in natural and artificial systems. Ann Arbor, University of Michigan Press, 1975

18. Goldberg David E. Genetic Algorithms in Search, Optimisation, and Machine Learning. Addison-Wesley Publishing Company, Inc., 1989

19. Serra R, Zanarini G. Complex systems and cognitive processes. Springer-Verlag, Heidelberg, 1990

# Solving Routing Problems with Cellular Automata

C. Hochberger, R. Hoffmann,
Microprogramming and Computer Architecture
Technical University of Darmstadt
Alexanderstr. 10, Darmstadt, D-64283, Germany

### Abstract

New variants of the classical Lee algorithm for the routing of connections on printed circuits or chips have been found when mapping it onto the cellular processing model. Two cellular algorithms are described: (1) a cellular shortest path algorithm with only 14 states per cell, which is independent of the grid size and (2) a parallel algorithm for routing multiple nets simultaneously.

## 1   Motivation

A large number of problems can be solved if they are described as cellular algorithms. Typical such problems are physical fields, lattice gas models, diffusion, hydrodynamics, wave optics, Ising spin systems, crystal growth, biological growth, games, artificial worlds, image processing, pattern recognition, simulation of digital circuits and graph algorithms.

The following routing algorithm was developed, when the expressive power of the language CDL[1][2] for the description of cellular algorithms was explored for different applications.

One of these applications was the problem of routing connections on a printed circuit board. It turned out, that by mapping classical algorithms onto the cellular model new or more efficient algorithms might be found. In this case, the classical Lee routing algorithm finding the shortest path between two points was transformed into an more efficient form, where only 14 states per cell are sufficient independently from the size of the grid.

The basic cellular routing algorithm can also be modified and extended in order to find multi-point connections or speed-up through parallelization.

## 2   Lee Algorithm

A very well known approach to routing problems is the Lee algorithm[7]. Its purpose is to find an optimal path between two points on a regular grid. Each point of the grid is associated with a weight. The algorithm finds the path on the grid with the lowest sum of weights. By adjusting the weights of the grid points the user has some control over what is supposed to be an optimal path. Let us consider an example: The user simply searches for the shortest path between two points. In this case the user specifies the weight one for all grid points and the algorithm will find the path with the lowest number of grid points. This is the shortest path between the two chosen points. In another example the user looks for a path that crosses the already existing paths as few as possible. In this case the user assigns a very high weight to all points of the existing paths and a very low weight to all other grid points. The algorithm will then find the path with as few crossings as possible.

The algorithm works in two phases. In the first phase the accumulated weights (acw) for each node relative to the starting point are computed. The accumulated weight for the starting point S is initialized to 0:

```
for all grid points i do
   acw(i) := infinity;
acw(starting point) :=0
steady := false
while not steady do
   steady := true;
   for all grid points i do
      min_neighbour := min(acw(neighbours(i)));
      acw(i) := min_neighbour+weight of this point;
      if acw(i) has changed then steady := false;
```

Figure 1 shows a sample grid in the middle of the calculation of the accumulated weights. The weight of all grid points is one in this case.

| $\infty$ | $\infty$ | $\infty$ | $\infty$ | $\infty$ | $\infty$ | $\infty$ | $\infty$ | $\infty$ | $\infty$ | $\infty$ |
|---|---|---|---|---|---|---|---|---|---|---|
| $\infty$ | $\infty$ | $\infty$ | $\infty$ | $\infty$ | $\infty$ | $\infty$ | $\infty$ | E | $\infty$ | $\infty$ |
| $\infty$ | $\infty$ | $\infty$ | 2 | $\infty$ | $\infty$ | $\infty$ | $\infty$ | $\infty$ | $\infty$ | $\infty$ |
| $\infty$ | $\infty$ | 2 | 1 | 2 | $\infty$ | $\infty$ | $\infty$ | $\infty$ | $\infty$ | $\infty$ |
| $\infty$ | 2 | 1 | S | 1 | 2 | $\infty$ | $\infty$ | $\infty$ | $\infty$ | $\infty$ |
| $\infty$ | $\infty$ | 2 | 1 | 2 | $\infty$ | $\infty$ | $\infty$ | $\infty$ | $\infty$ | $\infty$ |
| $\infty$ | $\infty$ | $\infty$ | 2 | $\infty$ | $\infty$ | $\infty$ | $\infty$ | $\infty$ | $\infty$ | $\infty$ |

**Figure 1**: Sample grid in the middle of phase 1

In the second phase the actual path is established. For this purpose the algorithm walks from the end point E towards the neighbour which has the smallest accumulated weight. This step is repeated until the algorithm reaches the starting point. Figure 2 shows a sample grid at the end of phase one. In figure 3 one of the possible resulting paths is shown. Note, that there are several possibilities to build a path from the end point to the start point. At the end point you can either go right or down.

| 7 | 6 | 5 | 4 | 5 | 6 | 7 | 8 | 9 | 10 | 11 |
|---|---|---|---|---|---|---|---|---|----|----|
| 6 | 5 | 4 | 3 | 4 | 5 | 6 | 7 | E | 9 | 10 |
| 5 | 4 | 3 | 2 | 3 | 4 | 5 | 6 | 7 | 8 | 9 |
| 4 | 3 | 2 | 1 | 2 | 3 | 4 | 5 | 6 | 7 | 8 |
| 3 | 2 | 1 | S | 1 | 2 | 3 | 4 | 5 | 6 | 7 |
| 4 | 3 | 2 | 1 | 2 | 3 | 4 | 5 | 6 | 7 | 8 |
| 5 | 4 | 3 | 2 | 3 | 4 | 5 | 6 | 7 | 8 | 9 |

**Figure 2**: Sample grid at the end of phase 1

| 7 | 6 | 5 | 4 | 5 | 6 | 7 | 8 | 9 | 10 | 11 |
|---|---|---|---|---|---|---|---|---|----|----|
| 6 | 5 | 4 | 3 | 4 | 5 | 6 | 7 | E | 9 | 10 |
| 5 | 4 | 3 | 2 | 3 | 4 | 5 | 6 | P | 8 | 9 |
| 4 | 3 | 2 | 1 | 2 | 3 | 4 | 5 | P | 7 | 8 |
| 3 | 2 | 1 | S | P | P | P | P | P | 6 | 7 |
| 4 | 3 | 2 | 1 | 2 | 3 | 4 | 5 | 6 | 7 | 8 |
| 5 | 4 | 3 | 2 | 3 | 4 | 5 | 6 | 7 | 8 | 9 |

**Figure 3**: One possible path

Obstacles can be modeled by infinite weights a grid points. Since the sum of the local weight and the accumulated weight of the neighbours will always be infinite, such grid points may never become part of the path.

At a first glance this algorithm looks like it perfectly fits onto a cellular automaton. Unfortunately the number of states required to perform the algorithm is related to the longest path or more precisely to the largest accumulated weight that may occur. Thus we decided to develop a version of the algorithm[5] which has a constant number of states. Our version can only handle the shortest path problem since it assumes a unified weight at all grid points.

# 3 Two point routing on cellular automata

The accumulated weights in the Lee algorithm are needed to find the shortest path. Instead of storing the accumulated weights we could also store the direction in which we have to move to get back to the starting point. With these marks instead of the accumulated weights the algorithm requires only a constant set of states. Of course we can not handle problems that have arbitrary weights on the grid points.

We present the modified algorithm in CDL. This language has been developed to describe cellular automata in a concise and readable way, independent of the target architecture. Compilers are available that translate CDL into C functions for almost any software architecture and into boolean equations for hardware simulators like the CEPRA family[3].

The algorithm can be described as follows:

```
(01)cellular automaton Lee_routing;
(02)
(03)const dimension=2;
(04)      distance=1;
(05)
(06)      cell=[0,0];
(07)      north=[0,1]; east=[1,0]; south=[0,-1]; west=[-1,0];
(08)
(09)type  celltype = (free,used,start,goal,
(10)                   /* phase 1: wave marks */
(11)                   wave_up,wave_right,wave_down,wave_left,
(12)                   /* phase 2: path directions */
(13)                   path_up,path_right,path_down,path_left,
(14)                   clear,ready);
(15)
(16)group wave = {wave_up,wave_right,wave_down,wave_left};
(17)      path = {path_up,path_right,path_down,path_left};
(18)      neighbours = {north, east, south, west};
(19)
(20)var   n : celladdress;
(21)      c : celltype;
(22)
(23)rule
(24) case *cell of
(25) free: if one(n in neighbours & c in wave: *n in {start,wave})
(26)      then *cell := c;
(27) goal: if one(n in neighbours & c in path : *n in {start,wave})
(28)      then *cell := c;
(29) wave: if (*north = path_down) or (*east = path_left) or
(30)         (*south = path_up) or (*west = path_right) then
(31)             *cell := element(celltype,value(*cell)+4)
(32)      else
(33)         if one(n in neighbours: *n in {path,clear}) then
(34)           *cell := clear;
(35) clear: *cell := free;
(36) start: if one(n in neighbours: *n in path) then
(37)             *cell := ready;
(38) end;  /* of case */
```

At the beginning of the first phase all cells are in the **free** state, except for the start and the end point. In the first phase all cells check whether there is any cell in the neighbourhood that already has a wave mark. If a wave mark is found, the cell itself becomes a wave mark towards the already marked cell. This is performed in line (25) and (26). The **one** function successively assigns all the elements of the groups **neighbours** and **wave** to the variables n and c. For each assignment the condition following the colon is checked. The evaluation of the **one** function stops if an assignment is found, that satisfies the condition, i.e. the corresponding neighbour is in state **start** or in any of the states **wave**. The first assignment is n=north and c=wave_up. The assignment to c is only used for its side effect to store the wave state corresponding to the neighbour being investigated. If the east neighbour is currently investigated this cell state will change to **wave_right** since it must point to this neighbour. Figure 4 shows a sample grid at the end of phase one. The wave marks are symbolized by small arrows. The black squares are obstacles. We had to introduce the special state **used** to model obstacles, since we do not have weights at the grid points.

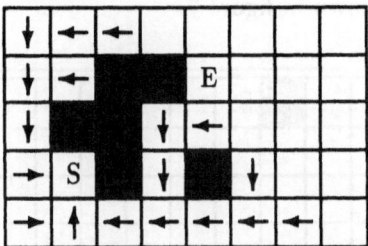

**Figure 4**: Wave marks at the end of phase one

Phase one ends when the end point (state **goal**, lines (27)-(28)) is reached by the wave. Now the path is built backward towards the start point along the wave marks (lines (29)-(34)). If a cell is one of the **wave** states and it sees a neighbour cell that is a path towards this cell, then this cell becomes a path in the direction of its previous mark (Fig. 5). This is done in the CDL program by adding four to the enumeration element. E.g. **wave_down** becomes **path_down**.

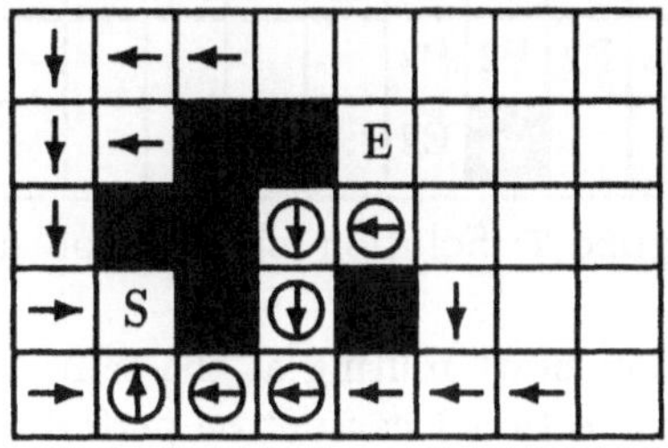

**Figure 5**: The path along the former marks

All wave marks from phase one have to be cleared in order to allow subsequent routing passes (lines (33)-(34)). For this purpose all cells that see a neighbour cell which is a path not pointing towards this cell are cleared. Such a cell will never become part of the path. Also all cells are cleared, that see a neighbour in the clear state. A cell that is in the **clear** state will become **free** in the next generation (line (35)). Since the building of the path moves along the shortest path, it is impossible that a cell in **clear** state can reach a cell which will be in the path but is not yet part of it.

The algorithm terminates when the **start** cell sees one if its neighbours become part of the path. The **start** cell changes its state to ready and thereby signals that the path is complete (lines (36)-(37)).

This implementation of our algorithm requires only 14 states and thus can very easily be realized in hardware.

# 4 Routing Multiple Nets Simultaneously

In order to make use of the inherent massive parallel computing power of a cellular automaton, it is desirable to route multiple connections simultaneously. Each single connection can be routed as in the simple case. Unfortunately some new problems arise in this case. Consider the situation in figure 6. Obviously a solution exists. It is shown in figure 7.

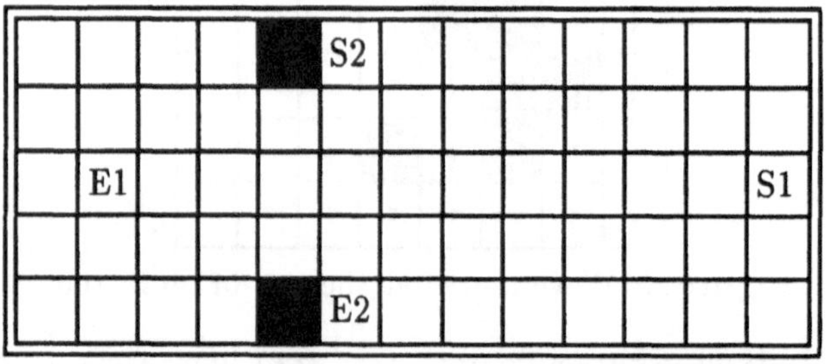

**Figure 6**: Two nets in a critical situation

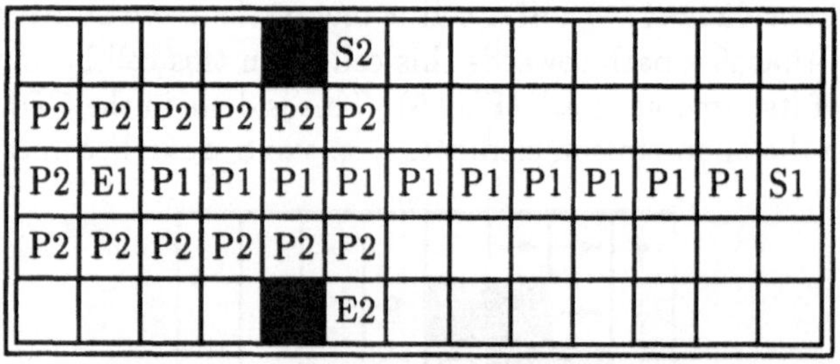

**Figure 7**: Solution for this problem

But if both connections start routing at the same time, connection 2 will form a barrier so that connection 1 can not be performed (Fig. 8).

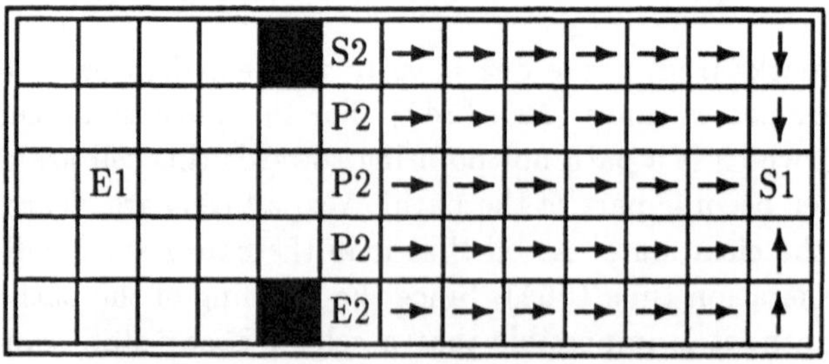

**Figure 8**: Short path blocks long path

The solution to this problem can be divided into two subproblems. In the first step a start point has to detect, that it can not establish a path to its end point. In the second step it has to remove paths that have already been formed, to be able to reach its end point.

How can a start point determine that there is no path to its end point? To solve this question let us first consider a single cell within the search wave. This cell can not become a part of the path if none of its neighbours points towards it. Figure 9 illustrates this situation. No neighbour points towards the marked cell. Even if one of the neighbours would become part of the path, the marked cell could never become part of the path. The marked cell can now change its state into `dead_end`. In the next generation the upper cell is in the same situation as the marked cell one generation earlier. This cell also can not become part of the path and thus changes its state to `dead_end`. This situation will happen in all vertices of the shown grid. The dead end information therefore will propagate towards the start point. As soon as all neighbours of the start point are in the state `dead_end` the start point knows that a path to the end point is impossible.

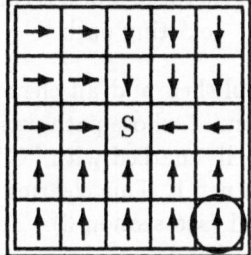

**Figure 9**: Wave mark, that can not become part of the path

The start point must be able to remove existing paths if they block the path to the end point. For this purpose we had to introduce a net priority. A search wave that has the same priority as an existing path will not cross that path. But if the search wave has a higher priority, then it will cross the path. In the beginning all nets start with the same priority. As soon as a start point detects that it can not find a path to the end point (Fig. 10), it increases its priority and spreads a new search wave. This new search wave will now cross existing paths with lower priority (Fig. 11).

|   |    |   |   |   | S2 | D | D | D | D | D | D | D |
|---|----|---|---|---|----|---|---|---|---|---|---|---|
|   |    |   |   |   | P2 | D | D | D | D | D | D | D |
|   | E1 |   |   |   | P2 | D | D | D | D | D | D | S1 |
|   |    |   |   |   | P2 | D | D | D | D | D | D | D |
|   |    |   |   |   | E2 | D | D | D | D | D | D | D |

**Figure 10**: Start point 1 detects deadlock

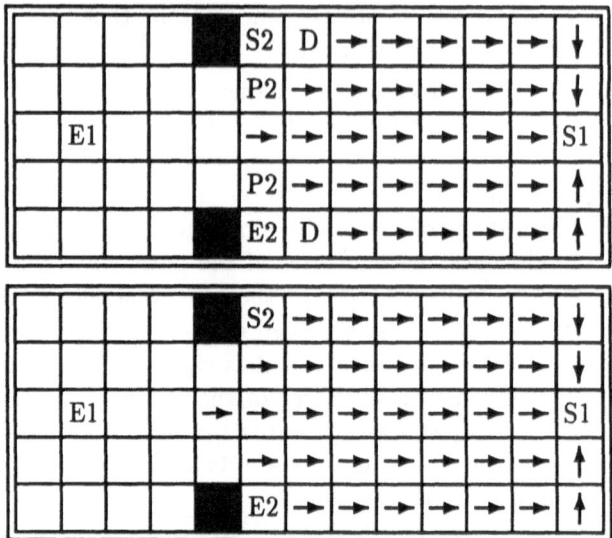

**Figure 11**: Two consecutive generations of a second search wave crossing the existing path

After the path has been established and all elements of the search wave are cleared from the grid, start point 2 whose path was broken will send out a new search wave (Fig. 12). The cells marked with "C" form a clearing wave that removes the wave marks from net 1. This clearing wave runs over the field as soon as the path is established (see section 3). The search wave of net 2 can only propagate over the field, after the wave marks of net1 have been cleared.

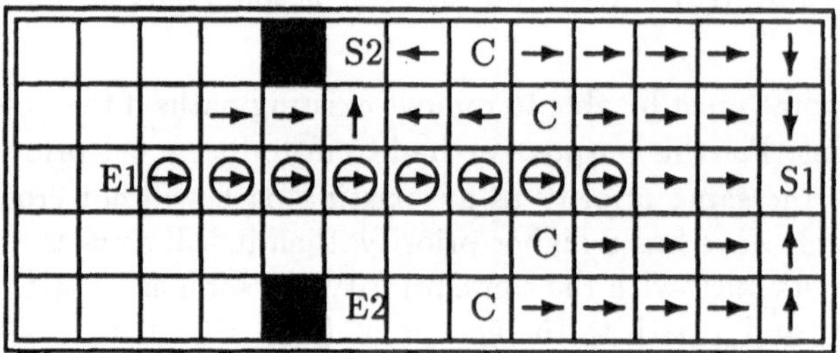

**Figure 12**: Start point 2 emits a new search wave

The presented algorithm requires a cell state with three components. The first component describes the state of the cell. In our implementation 24 discrete states are required. The second componenent stores to which net this cell belongs. Its complexity depends on the number of nets that should be simultaneously routed. The third component stores the priority of the net, to which this cell belongs. The number of priority levels should be related to the number of nets to be routed. An example with eight nets being routed in parallel and 4 priority levels would require $24 \times 8 \times 4 = 768$ states.

This algorithm still is not capable of solving all problems. If there are circular dependencies between more than two nets the algorithm may start to oscillate by successively increasing the priority of all the nets.

# 5 Open Problems

The two presented algorithms only deal with two point nets. A more realistic approach would require the possibility to solve multipoint problems. Thus we investigated algorithms for these problems. Unfortunately an optimal solution would require the construction of a Steiner tree[6][4], which is known as a NP problem. Thus our aim was to exploit the inherent parallelism of the cellular automaton instead of optimal solutions.

Different algorithms were implemented. The first one was exploiting the maximum parallelism trying to route all nodes of a net simultaneously except one. If all nodes are emitting a search wave a deadlock can occur. In this case the search waves of the different nodes were interfering with each other. This prevented the algorithm from finding good solutions. In a second version a considerable time was spent to reuse already routed paths. The results were generally better, but it took much more time than the first algorithm.

Both algorithms do not provide satisfying solutions up to now. Thus we are still trying to improve the algorithms.

Another problem that has not been addressed so far, are multilayered connections. Maybe this kind of problem can only be solved on three dimensional cellular automata.

# 6 Conclusion

By mapping classical algorithms onto the cellular processing model, more efficient and more parallel algorithms can be found. In this example, a modified version of the Lee routing algorithm was found, which is independent of the grid size and needs only 14 states (4 bits) per cell. Also innovative cellular algorithms have been found for multipoint connections and parallel routing. Generally speaking, the attempt of mapping classical algorithmic problems onto the cellular automata model might lead to new ideas in the theory and implementation of algorithms.

# References

[1] C. Hochberger and R. Hoffmann. CDL - a language for cellular processing. In Giacomo R. Sechi, editor, *Proceedings of the Second International Conference on Massively Parallel Computing Systems*, pages 41–64. IEEE, 1996.

[2] C. Hochberger, R. Hoffmann, and S. Waldschmidt. Compilation of CDL for different target architecures. In Viktor Malyshkin, editor, *Parallel Computing Technologies*, pages 169–179, Berlin, Heidelberg, 1995. Springer.

[3] R. Hoffmann, K.-P. Völkmann, and M. Sobolewski. The cellular processing machine CEPRA-8L. *Mathematical Research*, 81:179–188, 1994.

[4] X. Hong, T. Xue, E. S. Kuh, C.-K. Cheng, and J. Huang. Performance-driven steiner tree algorithms for global routing. In ACM-SIGDA and IEEE, editors, *Proceedings of the 30th ACM/IEEE Design Automation Conference*, pages 177–181. ACM Press, June 1993.

[5] Huschaam Hussain. Integration eines Compilers für die Zellularsprache CDL in das XCellsim–System. Master's thesis, Technische Hochschule Darmstadt, November 1994.

[6] M. Kaufmann, S. Gao, and K. Thulasiraman. On Steiner minimal trees in grid graphs and its application to VLSI routing. *Lecture Notes in Computer Science*, 834:351 ff, 1994.

[7] C.Y. Lee. An algorithm for path connections and its applications. *IRE Transactions on Electronic Computers*, pages 346–365, September 1961.

# Towards Multilayered Automata Networks

Stefania Bandini, Giancarlo Mauri
Department of Computer Science, University of Milan
Via Comelico 39 - 20135 Milan (Italy)
e.mail: bandini@dsi.unimi.it

## Abstract

In this paper Multilayered Automata Networks are formally defined as a generalization of Cellular Automata Networks.
They are hierarchically organized on the basis of nested-graphs, and can show different kinds of dynamics, which allow to use them to model, for example, complex biological systems comprised of different entities organized in a hierarchical framework.

## 1 Introduction

Multilayered Cellular Automata (also called Hyper-Cellular Automata) have recently been introduced as computational models for multi-level biological systems.

In particular they have been used to simulate cell interactions in the Immune System [1, 2], and the diffusion of Calcium ions inside cells [3, 4]. In the simulation of the Immune System and of Calcium diffusion two-level Multilayered Cellular Automata have been utilized.

In this paper we present a formal description of *Multilayered Automata Networks* defined as a generalization of *Automata Networks* [5], and the general features which characterize this particular computational model. A hierarchical structure is defined by means of a *nested-graph* [6], a graph composed by vertexes and arcs where each vertex is in turn a nested-graph. A nested structure is thus obtained.

The dimension of the nested-graph is given by the number of the nested graphs plus the first one, and represents the number of levels of the entire multilayered structure.

By introducing states and transition functions, a multilayered automata network is directly obtained from the nested-graph structure. The dynamical evolution of a multilayered automata network can be *sequential, synchronous* or *combined*.

Within this formal framework Multilayered Cellular Automata are introduced as a particular case of Multilayered Automata Networks where the highest level is a regular grid.

# 2 Nested Graphs

The definitions given in this section are the groundings of a complex general structure allowing different standpoints of a system description to be expressed. For instance, a multi-level structure as here introduced can be very useful to describe, at a given level, a biological element and, at a subsequent level, its constituent parts (e.g. cells), and, at another subsequent level, constituent elements (e.g., proteins or molecules), and so on.

## 2.1 Preliminary Definitions

*Definition 1*
A *graph* is a pair:

$$G = <V,l>$$

where
- $V$ is a finite or numerable set of elements called vertexes or nodes;
- $l:V\rightarrow \wp\ (V)$ is the neighborhood function, which determines for each node the set of adjacent nodes.

*Definition 2*
$G = <V,l>$ is locally finite if and only if $\forall v\in V\ |l(v)| < \infty$.

In the following, we will consider locally finite graphs.

The set of graphs will be denoted by $G$. The set $\mathcal{HG}$ of nested-graphs is recursively defined as follows.

*Definition 3*
- The nested-graphs of level $0$ are the graphs such that:

$$\mathcal{HG}_0 = G$$

- The nested-graphs of level $i+1$ are graphs whose nodes are mapped into nested-graphs of level $i$:

$$\mathcal{HG}_{i+1}=\{<G,\varphi> \mid G=<V,l>\in G,\ \varphi:V\rightarrow\mathcal{HG}_i\}$$

A nested-graph of level $i$ will be denoted by $HG^i = <G,\varphi>$. The graph $G$ will be called its *support graph*. A nested-graph is locally finite if and only if for each level $i$ the correspondent graph is locally finite.

Each node of the nested-graph is identified by a pair $<i,v>$ where $i$ denotes the level and $v$ is the name of the node in the graph of level $i$. The neighborhood of a node $<i,v>$ is defined by the neighborhood fuction $l^i$ related to the graph of that level. For sake of simplicity, we will use the notation $l_v$ instead of $l^i(v)$ when no ambiguity arises. Furthermore, we will have a nesting neighborhood function, $\delta$ which associates to every node $v \in V^i$ the set of nodes of the graph $\delta^i(v)$ at the level $i\text{-}1$.

As illustrated in Fig.1, let us consider a generic level $i$ of an extended graph $GE$. Both the level $0$ and the level $1$ are constituted by a family of graphs, and the level $2$ is composed by one connected graph consisting of two vertexes. If we take two vertexes of two distinct subgraphs of level $1$, there is no path from one to the other given by arcs of level $1$, but if the two subgraphs are adjacent vertexes of level $2$, then a path which connects them exists. It is given by the arc of level $2$ which link the two subgraphs. The same considerations hold for the vertexes of the subgraphs of level $0$, among which, although a path of arcs of level $0$ that connects them does not exist, a path composed by arcs of level $1$ or $2$ can exist.

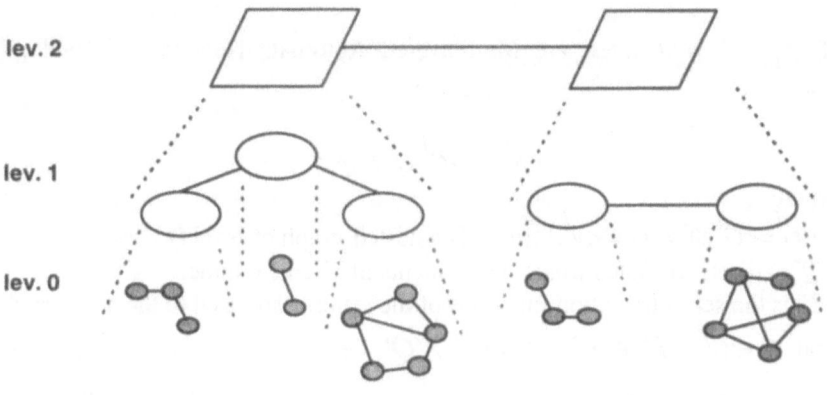

Fig.1: A three level Hypergraph $HG^2$

# 3 Multilayered Automata Networks

In this section we will give the definition of an extension of the Automata Network given in [5] called *Multilayered Automata Network*. This definition will be based on the above introduced nested-graphs: if we have a graph of level $k$; then the Automata Network will be composed by $k+1$ levels of nested networks. The state of a node at level $i$ can be modified as a function of the states of nodes in *i-th* neighborhood or as a function of the states of the nodes belonging to the structured nested automaton. In the particular case in which $k=0$, the Automata Network has only one level, giving a usual Automata Network.

In the following we will use the notation $q_v^i$ for denoting the state of the node $v$ of level $i$. More formally:

*Definition 4*

- A structure

$$A = <G, Q, F>$$

where
- $G = (V, l)$ is a graph;
- $Q$ is the finite set of states which the nodes of the graph can assume;
- $F = \{f_v | v \in V\}$, is the set of the transition functions of the states
such that $f_v : Q^{|l_v|} x Q \to Q$

is a Multilayered Automata Network of level $0$ and will be denoted by $A^0$;
the nodes $v \in V$ will be called nodes of level $0$ of the Automata Network $A$.

- If $A_k^{i-1}$, $k = 0, ..., n$ are Multilayered Automata Networks of level $i$-$1$, a structure

$$A^i = <HG^i, Q^i, F^i, g^i>$$

where
- $HG^i = <G^i, \varphi^i>$, $G^i = <V^i, l^i>$ (i.e. is a nested-graph of level $i$);
- $Q^i$ is the set of states which the elements of $V^i$ can assume;
- $F^i$ is the set of transition functions of the states associated to the nodes such that $\forall v \in V^i$, $f_v^i \in F^i$ $f_v^i : Q^{i|l_v^i|} \times Q^i \to Q^i$ ;
- $g^i : V^i \to Q^i$ such that $g^i(v) = g(q_1, ..., q_n) = q_e$, where $q_j$ $(j=1,...,n)$ is the state of the node $v_j^{i-1} \in \delta(v)$ and $g' : Q^{i - |V^{i-1}|} \to Q^i$

is a Multilayered Automata Network of level $i$. The Automata Networks $A_k^{i-1}$ are called nodes of level $i$ of the Automata Network $A^i$.

# 4 The Dynamics

For the biological systems we faced so far, a combined procedure appeared the most adequate, where updating by the synchronous application of reaction rules is alternated to updating by applying diffusion rules.

Let us introduce for the Multilayered Automata Networks the equations which describe the network dynamics, i.e. the sequence of the states depending on time.
Let us consider a totally ordered discrete set of temporal steps $T = \{t_0, t_1, ...\}$ where $t_0$ is the minimum element. Such a set represents the time steps for observing the evolution of the states.

As for the Automata Networks, it is possible to give different types of dynamics also for the Mulatilayered Automata Network, depending on the fact that the state transition occurs in parallel in all the nodes of the network or in a sequential way.

It is possible to have *synchronous* or *sequential* iterations, or a combination of both among the nodes of a same level or among different levels.

We denote by $q_v^i(t)$ the state of the node $v \in V^i$ at the instant $t$ and by $q(t)$ the global configuration of the system at $t$.

### Synchronous Iteration

The *synchronous* iteration of a network $A^k = <G^k, Q^k, F^k, g^k>$ is actived when all the states of all the nodes of the network are updated in parallel.

The equation for the levels $i$ with $i > 0$ is:

$\forall i=1,...,K, \forall v \in V^i$

$$q_v^i(t+1) = f_v^i(q_p^i(t), q_e(t)) \ (p \in l_v^i)$$

where $q_e \in Q^i$, $q_e = g^i(v)$ and for the level $i=0$

$$q_v^0(t+1) = f_v^0(q_p^0(t), q_v(t)) \ (p \in l_v^0).$$

### Sequential Iteration

In the *sequential* iteration the states of the nodes are updated one at a time, by levels from $k$ to $0$ and, at every level, according to a totally ordered relation $\leq$ defined at each level of the network.

The related equation for the levels $i$ with $i > 0$ is:

$\forall i=1,...,K, \forall v \in V^i$

$$q_v^i(t+1) = f_v^i(y_p^i, q_e(t)) \ (p \in l_v^i)$$

where

$$y_p^i = \begin{cases} q_p^i(t+1) & \text{if } p < v \\ q_p^i(t) & \text{if } p \geq v \end{cases}$$

where $q_e \in Q^i$, $q_e = g^i(v)$ and for the level $i=0$

$$q_v^0(t+1) = f_v^0(y_p^0) \ (p \in l_v^0)$$

where

$$y_p^0 = \begin{cases} q_p^0(t+1) & \text{if } p < v \\ q_p^0(t) & \text{if } p \geq v \end{cases}$$

Notice that in this type of iteration the nodes are updated in a sequential way both in respect to the levels an to the order defined on each level, i.e., the state transition occurs in the order given on the nodes of the level $k$, then on the nodes of the level $k+1$ and so on.

*Combined Iteration*

One of the most useful applications of this kind of dynamic is rappresentated by the opportunity to model reaction-diffusion systems, as one can see in [7], where a combined dynamics is used for the specification of multilayered automata networks as multilayred cellular automata, as defined in the next section.

In the case of the Multilayered Automata Networks, it is possible to have a dynamics where the levels are updated in a sequential way, but where at each level the updating is synchronous.

The equation for the levels $i$ with $i > 0$ is:

$$\forall i=1,...,K, \ \forall v \in V^i$$

$$q_v^i(t+1) = f_v^i(q_p^i, q_e(t)) \ (p \in l_v^i)$$

where $q_e \in Q^i$, $q_e = g^i(v)$

and for the level $i=0$

$$q_v^0(t+1) = f_v^0(q_p^0) \ (p \in l_v^0).$$

# 5 Hypercellular Automata

A two level multilayered cellular automaton is comprized of a first level, where a two-dimensional cellular space represents the *diffusion space*, and of a second level where a totally connected graph corresponds to each cell to reproduce the *reaction space*. Updating of the two level multilayered cellular automata involves two main classes of rules:

- *reaction rules*: these rules dictate state changes. The state of each entity (*actor*) in a first level site is the set of values denoting specific attributes associated to it in order to let it play the role of a specific type of biological entity; such state changes after viewing the states of all the other adjacent entities in the same first level site. These rules are active at the second level, constituted by totally connected graphs.

-*diffusion rules*: these rules determine the diffusion of entities on the cellular space. They operate at the first level to displace the entities from the site where they are momentarily located to one of the neighboring sites.

*Definition 5*
A graph $G = (V,l)$ is a cellular space iff

*1.* $V = \mathbf{Z}^d$;

2. $\forall k \in \mathbf{Z}^d$, $\forall i, j \in V$ $i \in l(i) \leftrightarrow j+k \in l(i+k)$.

A Multilayered automata network of order $0$ where $G$ is a cellular space with a sincronous dynamics is a *Cellular Automata*.

## References

1. Celada F, Seiden P E. A Computer Model of Cellular Interaction in the Immune System. Immunology Today 1992; vol. 13, no. 2
2. Bandini S. Hyper-Cellular Automata for the Simulation of Complex Biological Systems: a Model for The Immune System. In Kirschner D (ed) Special issue on "Advances in Mathematical Modeling of Biological Processes". Int. Journal of Applied Science and Computation 1996; vol. 3, no. 1, ISSN 1076-5131
3. Bandini S, Fesce R, Mauri G. Multilayered Cellular Automata in Modeling Biological Systems. Proc. 3rd Systems Science European Congress, Rome, 1996
4. Bandini S, Malagutti G, Fesce R. A Composite Automaton Designed to Model Complex Biological Systems. These proceedings.
5. Goles E, Martinez S. Neural and Automata Networks: Dynamical Behavior and Applications. Kluwer Academic Publishers, Boston, 1990
6. Sieburg H B. The Cellular Device Machine: Point of Departure for Large-Scale Simulations of Complex Biological Systems. Computers Math. Applic. 1990. vol.20, no. 4-6; 247-267
7. Bandini S, Bogni D, Tarantello G. A Cellular Automata Approach To Reaction-Diffusion Models: A Proposal. Proc. of the First National Conference on "Automi Cellulari nella Ricerca e nell'Industria", (ACRI '94). Rende, 1994
8. Langton C G. Artificial Life. Addison Wesley, New York, 1989
9. Pozzan T, Rizzuto R, Volpe P, Meldolesi J. Molecular and Cellular Physiology of Intracellular Calcium Stores. Physiological Reviews 74, 1994
10. Rosen R. Dynamical Systems Theory in Biology. John Wiley and Sons Inc., 1970
11. Serra R, Zanarini G. Complex Systems and Cognitive Processes. Springer-Verlag, 1992
12. Turing A. The Chemical Basis for Morphogenesis. Philos. Trans. R. Soc., London, 1952.

# A COMPOSITE AUTOMATON DESIGNED TO MODEL COMPLEX BIOLOGICAL SYSTEMS. Modeling the calcium-ion distribution in the living cell

S. Bandini(+), R. Fesce(*), G. Malagutti (+*)

(+)Dipartimento di Scienze dell'Informazione
Università di Milano
via Comelico, 39 - 20134 Milano (ITALIA)
tel. +39-2-55006269  fax +39-2-55006276
e.mail: bandini@dsi.unimi.it

(*)CNR Centre Cellular Molecular Pharmacology
DIBIT Centre Theoretical Biology, HSR - Milano (ITALY)
tel .+39 -2 26434809, fax +39 -2 26434813
e.mail: fesced@dibit.hsr.it

## Abstract

A composite automata endowed with the properties of parallel and local processing of information by means of mutual interactions between neighboring elements, is proposed as an appropriate framework to build efficient and accurate models of biological systems. It is based on the hypercellular automata model and is characterized by three main features in addition to the basic properties of it: collective actors; cellular space and compartments; a main set of molecular reaction and diffusion rules, and two further sets of rules governing collective reaction and diffusion. The purpose of this work is to describe the implementation of a composite automaton, which has evolved from hypercellular-automata models of reaction-diffusion systems, to model the distribution of calcium ions ($Ca^{2+}$) in the subcompartments of a living cell.

## 1. General problems in modeling biological systems

The general properties of biological systems arise from complex interactions of lower level subsystems, which can in turn be decomposed, down to the stochastic interactions among single molecules. The first task in modeling biological systems is therefore to define the "viewpoint", i.e. the degree of detail required for an adequate description of the system.

In particular, if modeling is aimed at gathering some insight into the mechanisms which underlie the process of interest — rather than merely reproducing the observed behaviour — the "viewpoint" will often have to be multiple, i.e. it will be necessary to simulate various aspects and subsystems at different degrees of detail and with different temporal and/or spatial resolution. Further problems often arise in defining the simulation space, as the topological features of biological systems are in many cases complex and heterogeneous.

These observations suggest that composite automata, endowed with the properties of parallel and local processing of information by means of mutual interactions between neighboring elements, may constitute the most appropriate framework to build efficient and accurate models of biological systems.

Cellular automata are the best characterized local and parallel computational models. However, cellular automata in their classical form were not designed to cope with topological heterogeneity and with the coexistence of multiple subsystems and different "particle" populations. In classical applications of cellular automata, such as fluid dynamics, the cellular space [8] [10] is homogeneous, one or a few types of particles exist, interactions among particles are mostly limited to collision, and neighborhood rules govern the movement of particles. Modeling biological systems willl require introducing the representation of several different types of elements and the specific reactions among such elements, as well as the differential diffusion of the mobile species. To this purpose, it can be convenient to split the two aspects of *reaction* and *diffusion*, moving to more complex computational models (an example within a classical framework of cellular automata can be found in [5]).

The concept of "hypercellular automata", or multilayered cellular automata, has recently been proposed by [1] and [2], as a particular case of multilayered automata network. A hierarchical structure is defined through a *nested graph* [2], i.e. a graph composed by vertices and arcs, where each vertex is in turn a nested graph. The multilayered automata network is directly obtained from this structure by introducing status attributes and transition functions. Two-level multilayered cellular automata have been developed and employed to model biological systems: the first level constitutes a bidimensional cellular space (*diffusion space*), while at the second level a totally connected graph corresponds to each cell (first-level vertex) to generate an intrinsically parallel and local *reaction space*.

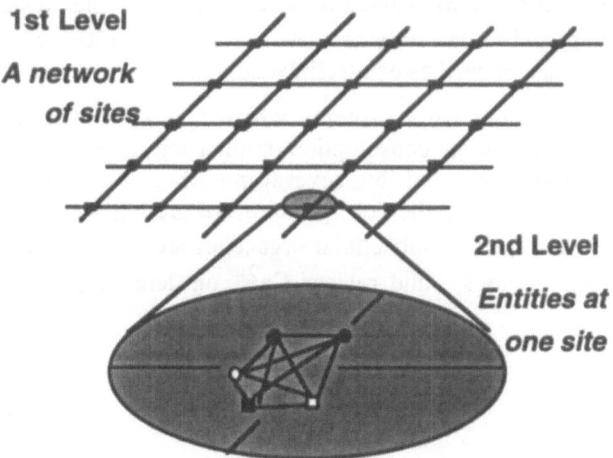

*Figure 1. The hypercellular automaton*

The dynamical evolution of a multilayered automata network can be sequential, synchronous or combined. For the biological systems faced so far [2], a combined procedure appeared the most convenient, i.e. updating by the alternate synchronous application of reaction rules or of diffusion rules. *Reaction rules* dictate status changes. The status of each entity (*actor*) is the set of values denoting specific attributes associated to it in order to determine its properties and behaviour; such status is changed (*reaction*) based on the status of all the other entities momentarily present in the same cell; thus, these rules apply to the totally connected graphs which correspond, at the second level of the nested graph, to single first level sites. *Diffusion rules* determine the diffusion of entities on the cellular space. They operate on the first-level cellular space to possibly displace the entities from the site where they are momentarily located to one of the neighboring sites.

Modeling a complex biological system entails a series of further complications, such as heterogeneous topological domains (*compartments*), differential constraints to diffusion, delays in the response of specific agents or subsystems to specific stimuli, and the need to simulate different aspects (or subsystems) with different temporal/spatial resolution.

The purpose of this work is to describe the implementation of a composite automaton, which has evolved from hypercellular-automata models of reaction-diffusion systems, to model the distribution of calcium ions ($Ca^{2+}$) in the subcompartments of a living cell.

## 2. Modeling calcium ion distribution in the cell.

Many biochemical processes in living cells are regulated by the concentration of free calcium ions ($Ca^{2+}$), from contraction of muscle fibres to secretion, proliferation and cellular death. Such processes are activated by widely different concentrations of $Ca^{2+}$, from fractions to hundreds of micromoles per litre (mM). Thus, living cells have developed powerful and sophisticated means to maintain very low basal calcium concentrations (about 0.1 mM, versus a more than $10^4$ higher concentration in extracellular fluids) and to carefully regulate elevations in $Ca^{2+}$ concentration in response to intracellular as well as extracellular stimuli.

Regulatory systems range from proteins which "pump" out calcium ions by coupling this process to energy consumption or other ion fluxes, to proteic pores or "channels" — finely regulated by several possible mechanisms — in the plasmamembrane surrounding the cell; proteins are present which can bind huge amounts of $Ca^{2+}$, and specific subcellular organelles are designed to actively store (by means of other "pumps") and release $Ca^{2+}$ on demand (by means of other "channels").

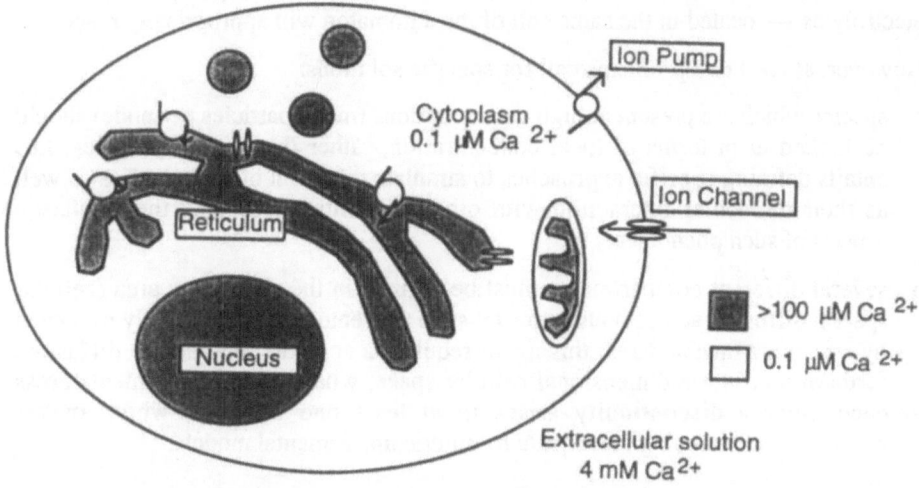

*Figure 2. Schematic diagram of the subcompartments in a living cell, with the main Calcium-ion transport mechanisms across biomembranes*

Such a multifarious and complex regulation, over more than 3 decades of $Ca^{2+}$ concentration, raises some modeling problems:

o  a classical representation of calcium ions as single particles would require up to about $10^6$ particles per $mm^3$ of simulated cell volume (the size of an average cell is about 500 $mm^3$), and a lattice-gas-type automaton would require about as many nodes (over 10 million for a two-dimension thin section of an average cell); furthermore, to obtain a reasonable diffusion speed for particles the simulation time step should be in the order of microseconds; on the other hand, molecular simulation of calcium ions is somehow required to reproduce stochastic fluctuations in calcium-regulated processes

o  the $Ca^{2+}$ regulation system, like most biological systems, is comprized of multiple biochemical subsystems where many different elements interact in various ways; therefore, the model must be able to host several different species of "particles", and each such element must play its role as an actor on stage

o  for many sub-systems, rates and modulatory mechanisms are reasonably well characterized; therefore, although this is in contradiction with the need for low-level molecular simulation, it would be very convenient to directly introduce such knowledge as high-level rules in the simulation system.

Multilayered cellular automata offer an effective approach to the question. Once appropriate time and space scales are set, then molecules present in low concentration will be allowed to move around according to standard rules of simulated diffusion; all molecules — proteins, channels, pumps as well as electrolytes — located in the same cell of the automaton will appropriately react.

However, at least two problems call for specific solutions:

o   species which are present at high concentrations (many particles per node) should be looked at in terms of local concentration, rather than single particles; this entails defining specific approaches to simulate diffusion of such species as well as their molecular interaction with other elements, preserving the stochastic aspects of such phenomena

o   several different compartments must be defined in the simulation area (cellular space); diffusion across boundaries must be prevented, or differentially regulated by *transport* mechanisms; this again requires a specific approach to diffusion; furthermore, in a bidimensional cellular space, where two compartments cross each other a discontinuity arises in at least one of them, which makes bidimensional mapping inadequate for multicompartmental models.

## 3. The composite automaton

In order to tackle these challenges, a composite automaton has been implemented by introducing three main features in addition to the properties described above for two-layered reaction-diffusion hypercellular automata.

### a. Collective actors

The composite automaton is based on a cellular space simulating the living cell; objects endowed with status attributes simulate chemical substances present in the living cell, and will be called "actors"; actors can move to neighboring cells according to *diffusion* rules and can react with other actors momentarily present in the same cell according to reaction rules (which change status attributes). Calcium ions, which may be present at high concentrations (up to >100 per cell), are not considered singularly. Rather, *collective actors* are introduced in the automaton; such collective actors have a fixed location, are present in all cells where $Ca^{2+}$ can occur, and have a status attribute dedicated to represent the number of Ca ions momentarily present in the cell. Single Ca ions are represented by bits in a fixed-length string, the *concentration* attribute, which is randomly scrambled at each step in the life of the system; $Ca^{2+}$ concentration will intervene in any reaction which is influenced by $Ca^{2+}$, by means of string-matching operations on the *concentration* string. This ensures that $Ca^{2+}$-dependence of any process is modeled even in its stochastic aspects. Thus, the function of collective actors is to increase the resolution of the model down to the molecular level, without concurrently requiring shorter time steps and a finer spatial grid; they constitute the "underground" portion of the model, working like lower-level automata.

### b. Cellular space and compartments

The existence of different compartments is handled by assigning each element in the cellular space to a specific compartment (e.g. extracellular, cytoplasm, intracellular organelles, endoplasmic reticulum). Specific status fields will determine which compartments each *actor* can reside in. In general, barriers to diffusion will exist at boundaries between cells belonging to different compartments; however, specific status attributes may endow actors with the capacity, or possibly the duty, of crossing such boundaries. Conversely, diffusion will generally be free among cells belonging to the same compartments.

All this profoundly affects the fundamental properties of the cellular space, which becomes a graph where different kinds of vertices and arcs are allowed (i.e. different compartments and direct connection or different boundaries between cells).

A further generalization with respect to classical cellular space is introduced in the automaton to cope with the limitations arising from a bidimensional simulation area: the cellular space is created as a graph where each vertex (cell) is connected to *at least* four other vertices, by means of arcs of possibly different kinds (depending on the compartments the two connected cells belong to). A subpopulation of the vertices and arcs map, respectively, to cells in a bidimensional simulation area and to the neighborhood relations among such cells; this makes it possible to display a representation of the simulation area. The other vertices map to cells outside the bidimensional simulation area and provide — together with the arcs originating from them — (i) the continuity when two compartments cross each other and (ii) a virtual extension of the bidimensional simulation area where this has to be arbitrarily truncated.

*c. The rules*

In addition to the main sets of rules mentioned above, i.e. reaction and diffusion rules, which we may more appropriately call *molecular* reaction/diffusion rules, the automaton contains two further sets of rules, which govern *collective* reaction and diffusion.

o   *Molecular diffusion rules* apply to all *mobile* actors; they govern the movement from a site to one of the neighboring sites (be they inside or outside the displayable bidimensional area). For each actor in each location, the applicable rules are directly determined by the status attributes of the actors, by the type of the cell and by its neighborhood relations (i.e. by the type of the corresponding vertex and of the arcs originating from it): in other words, the diffusion rule *A-C-C'* determines how an actor with diffusion attribute *A* which is located in a cell of a specific compartment *C* can move to a neighboring cell, belonging to compartment *C'*.

o   *Reaction rules* apply to all actors, simple as well as *collective* ones; reactions occur according to a totally-connected-graph scheme, i.e. they involve, for each cell, all actors momentarily present that particular cell. However, reaction fields in the structure of each actor determine, by means of string-matching rules, the other actors they can significantly interact with. Object oriented programming makes it easy to create *methods* for each actor which produce the desired changes in status attributes following recognition of a matching actor. Self reaction is also allowed to reproduce delays (countdown in a specific field) and complex responses (stepping through a series of states).

o   *Collective reaction rules* are just a subclass of reaction rules, with the exception that they do not apply to the totally connected graph of actors simultaneously present in the same cell, but rather on a reduced graph where each *collective actor* is connected to all other actors in the cell which possess the appropriate matching string. These rules are generally more complex than *molecular* reaction rules in that string matching of the *concentration attribute* to appropriate strings presented by the counterpart can determine graded changes in specific status attributes, thereby simulating graded activation/inactivation of proteins and mechanisms as functions of $Ca^{2+}$ concentration, preserving in the mean time the stochastic aspects of molecular interactions. Reactions between collective actors are allowed to simulate, for example, the binding/unbinding of calcium-ions to $Ca^{2+}$-buffering proteins, through the comparison of the concentration strings of collective actors which respectively represent the populations of free and bound calcium ions in the cell. This simulates slowing down of $Ca^{2+}$ diffusion by the buffering proteins.

o *Collective diffusion rules* apply to collective actors only, and in particular to collective actors located in adjacent cells. Adjacency between two cells for collective diffusion is determined by the existence of permissive arcs between the two vertices that represent the two cells, in the graph which defines the simulation space. These rules consist in the swapping of variable length substrings between the *concentration attributes* of the two involved collective actors, and simulate the movement (diffusion) of particles present at high concentration, by means of changes in status attributes, rather than location, of specific actors. Thus, as we shall shortly see, these rules must be formally distinguished from both reaction and molecular diffusion rules.

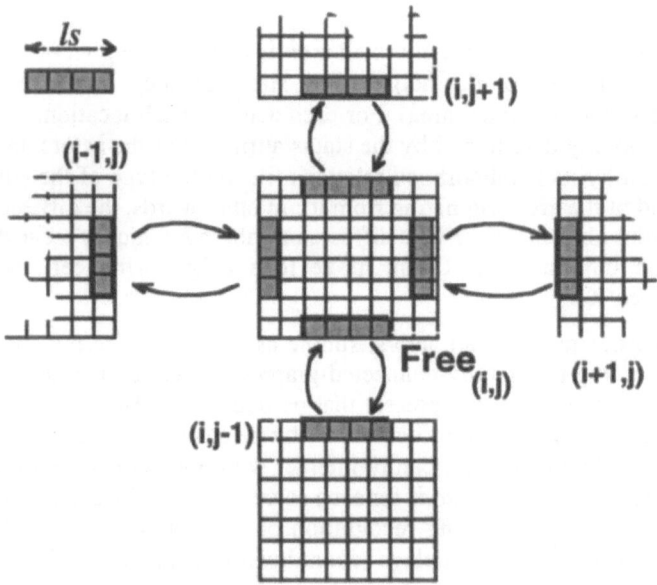

*Figure 3. $Ca^{2+}$ diffusion by swapping concentration attribute substrings*

Let us indicate by $D_M$ and $D_C$ the procedures to update the system by applying the *molecular diffusion* rules and the *collective diffusion* rules, respectively, and by $R_M$ and $R_C$ the procedure to update the system by applying the *molecular and collective reaction* rules, respectively. Let us also indicate by $G_0$ the graph where each vertex correspnds to a cell in the simulation space and the arcs define the neighborhood relations, and by $G_i$ {i=1,N} — where $N$ is the total number of vertices of $G_0$ — the totally connected graph dynamically generated at each step by the set of actors momentarily present in the cell mapped by vertex $i$ of $G_0$.

We can observe that $R_M$ and $R_C$ are formally similar, in that they apply to $G_i$ {i=1,N} and work independently for each cell, i.e. the execution of these procedures on the whole system is exactly equivalent to the parallel execution of the same rules on each single cell of the automaton. Furthermore, the execution of $R_m$ and $R_c$ do not produce any displacement of actors in the simulation space but only changes in the status attributes.

Conversely, execution of the $D_M$ procedure is exactly equivalent to the parallel application of molecular diffusion rules to each single mobile actor. This procedure brings about changes in the location of actors in the simulation space but does not affect any status attribute.

This reproduces the two-step updating procedure of a reaction-diffusion hypercellular automaton, where the two steps apply to two different domains — i.e. the cohort of actors for the diffusion step and the set of cells for the reaction step.

Execution of the $D_C$ procedure, finally, applies to a third, different domain, because it is equivalent to the parallel application of the corresponding rules to each *arc* in $G_0$. Furthermore, it simulates the diffusion of elements, present at high concentration in the system, by affecting the status of collective actors, i.e. it simulates diffusion but is formally equivalent to a reaction step (albeit the reacting actors are present in adjacent cells rather than in the same one).

## 4. Notes on the implementation

*a) The user interface*

The model is set up by the user who defines $G_0$, starting from a graphical representation of the bidimensional displayable space (possibly mapped on a digitized micrograph obtained experimentally from a living cell) and adding "external" cells where needed to grant the continuity of intersecting compartments.

The initial population of actors is generated and their initial location and set of status attributes are defined. This procedure can be performed, at will, either on the whole simulation area (independently for each compartment) or on selected regions or single cells. The parameters of most reactions can be set to default values or tuned at will.

When the simulation starts, a window displays the simulation area: a color code represents local calcium concentrations and the location of selected actors can be displayed. A second window yields full information on the actors momentarily present in a chosen cell, and lets the user modify their attributes. A control window lets the user suspend/restart the execution, select the monitored area and cell, change the number of actors and define the mode for data saving: time courses of relevant status attributes of specific actors, time courses of average $Ca^{2+}$ concentration and transcompartmental fluxes, in the selected area, can be saved; images of local $Ca^{2+}$ concentration can be saved at the desired time intervals.

*b) The actors*

Actors are present in the model to represent each relevant biological aspect. An actor does not necessarily correspond to a biological entity (see, below, collective actors and *shuttles*).

Each actor is a software object containing (i) a series of status attributes, (ii) a binary string which determines, by means of boolean operations, which other actors it can interact with, and (iii) all the procedures aimed at changing its status attributes as a consequence of the interaction with other actors (*reaction rules*). Most status fields regard the "functional state" of the actor, and its behavior in *reaction* procedures. Diffusive behavior is affected by an *access* field which defines the compartments the actor can reside in and move to, and by a *direction* fields which determines whether the actor is fixed or, in case it is mobile, whether diffusion is random or along an assigned direction (this permits the deterministic translocation of actors).

The main classes of actors are *collective actors*, *carrier actors* and *mobile actors*.

Collective actors — called *Free* and *Bound* — are fixed and represent calcium concentrations in the cell they are located in, by means of a binary string — the *concentration string* — where each set bit represents a fixed number of calcium atoms (1 in the cytoplasm, about 100 in compartments where calcium concentrations are higher). $Ca^{2+}$ diffusion within each cell is simulated by shuffling *Free*'s binary string, $Ca^{2+}$ binding/unbinding is simulated by transferring bits from suitable length substrings in *Free* to *Bound* and viceversa, and diffusion to neighboring cells is simulated by swapping fixed length substrings with *Free* actors in the neighboring cells. The occupancy of $Ca^{2+}$ receptors and the degree of activation of $Ca^{2+}$-dependent processes are easily defined by reaction rules which look up the number of set bits in substrings from the concentration string of either *Free* or *Bound*.

Carrier actors simulate $Ca^{2+}$ channels and pumps. They are located in cytoplasmic cells confining with other compartments, and they are in particular associated with a single arc in $G_0$ which connects two vertices of different kind. These actors simulate the opening of a pore or the active transport of calcium ions in a graded fashion, depending on their status. This allows for any kind of complex regulation of such mechanisms. The actual translocation of calcium ions is performed by *shuttle* actors, which are generated with the appropriate capacity and released into the cell; *shuttles* are loaded with $Ca^{2+}$ by reacting with *Free* and mandatorily migrate along the appropriate arc of $G_0$ for a defined number of steps (usually 1 or 2, i.e. back and forth).

Mobile actors simulate any diffusible biological entity of interest — mostly regulatory molecules — which can be generated by specific reactions and in turn react with specific actors. They usually have a *life* field in their status which limits in time their persistence in the automaton. They may exhibit *access* restrictions to the different compartments and mqy diffuse either at random or along obligatory directions (see the description of the *shuttles* above).

*c) Algorithms and boolean operations on status strings*

A series of algorithms have been developed to appropriately simulate diffusion and complex dependence on $Ca^{2+}$ concentration. Based on experimentally determined chemical-physical parameters and on theoretical computations, the probability of a particle leaving a fixed-volume cell during a fixed-duration time step in any direction is computed, thereby fixing the probability for a mobile actor to be displaced from its current location and the extent of the substring in *Free*'s concentration attribute to be swapped with neighboring *Frees*. Based on the knowledge of the affinity for $Ca^{2+}$ of the binding proteins and of other $Ca^{2+}$ receptors, the momentary occupancy of the latter is determined by boolean operations on substrings of *Bound*'s concentration attribute. Boolean operations on substrings of *Free*'s concentration attribute also reproduce polynomial approximations to high-order functions that describe the dependence of specific functions on $Ca^{2+}$ concentration.

# Conclusion

The model here described to simulate the dynamic regulation of $Ca^{2+}$ distribution in the subcompartments of a living cell constitutes a novel implementation of a composite computational automaton, based on an extension of reaction-diffusion hypercellular automata.

The basic structure of a cellular automaton for the simulation of diffusion is extended by introducing different kinds of particles and a second layer of totally connected graphs to simulate local and parallel reaction among the particles (hypercellular automaton).

The hypercellular automaton, where diffusion rules operate on the vertices of lower-level graph (cells) and reaction rules operate on the corresponding set of second-level graphs, is further extended by introducing (i) the existence of different kinds of cells and restricted diffusion in the first-level graph, (ii) a set of updating rules that operate on the arcs of the first-level graph and (iii) particles that are *de facto* automata themselves. The latter make it possible to reproduce underlying molecular processes whose simulation would require a finer spatial/temporal resolution.

The general features of this model permit to tackle the simulation of the particularly complex regulation of $Ca^{2+}$ distribution in the living cell. Most problems here discussed also occur in their general features in other complex biological systems, where many different agents are present, several compartments must be considered, different degrees of detail are required in the simulation of different aspects, and high order functions describe the dependence of certain functions on regulatory factors. It appears therefore that the composite automaton here presented offers a general scheme which should make many apparently untractable biological processes amenable to computer simulation.

## Reference List

1. Bandini S. Hyper-cellular Automata for the Simulation of Complex Biological Systems: a Model for The Immune System. In: J. of Applied Science and Computations 1996; vol.3, n.1, ISSN 1076-5131.

2. Bandini S, Mauri G. Towards Multilayered Cellular Automata, see this volume.

3. Celada F. La Logica della Risposta Immunitaria. In: Celada F (ed.) La Nuova Immunologia. Le Scienze Editore, Milano, 1992.

4. Celada F, Seiden P E. A Computer Model of Cellular Interactions in the Immune System, vol.13, n.2, Immunology Today. Elsevier, 1992.

5. Dab D, Boon J P. Approach to Reaction Diffusion Problems. In: Manneville P, Boccara N, Vichniac G Y, Bidaux R (eds.) Cellular Automata and Modeling of Complex Physical Systems, Springer Verlag, 1990.

6. Farmer J D, Packard N H. The Immune System, Adaptation and Machine Learning. Phisica 22D, 1986.

7. Frish, U., Hasslacher, B., Pomeau, Y., *Lattice-Gas Automata for the Navier-Stokes Equation*, Physical Review Letters, vol. 56 n.14, 1986.

8. Goles, E., S. Martinez, *Neural and Automata Networks: Dynamical Behavior and Applications*, Kluwer Academic Pu., Boston, 1990.

9. Pozzan, T., R. Rizzuto, P. Volpe, J. Meldolesi, *Molecular and Cellular Physiology of Intracellular Calcium Stores*, Physiological Reviews 74, 1994.

10. Wolfram, S., *Theory and Applications of Cellular Automata*, World Scientific, Singapore, 1986.

# Time Dependent Grid Simulation by a Cellular Automata Network

Dario Castagnolo*, Mario Mango Furnari**,
Francesco Mele*, Renata Napolitano**

* MARS Center
Napoli, Italy

**Istituto di Cibernetica, CNR
Napoli, Italy

## Abstract

The simulation of certain physical phenomena requires a degree of accuracy that varies according to the region of the physical system that needs to be modelled.

The time dependent grid method, which is utilised in the traditional numerical simulations is a good approach for confronting this problem.

In constructing models for physical phenomena that make use of this method, the current work aims at clarifying certain aspects regarding the methodology. We have made use of CANL (Cellular Automata Network Language) to represent the physical models by means of a composition of well distinguished model components. We have tested the time dependent grid simulation approach with cellular automata network applying it to the phenomenon of heat diffusion with phase change.

We report the results of this simulation, that consist in temperature patterns and relative grids at some relevant time steps.

## 1 Introduction

In this paper we propose to use a method based on Cellular Automata for modelling the computational approach of the time dependent grid.

To this end we have analysed the problem of heat diffusion with phase change; the corresponding numerical solution is based on a grid that becomes finer according to the location of the mushy region. The mushy region is the zone where the phase change occurs; here the physical properties (specific heat, conductivity) show a discontinuous jump. Therefore, from a mathematical point of view it is necessary to model this jump by means of appropriate interpolating polynomials. Also, from an energy balance point of view it is necessary to satisfy the conservation law; this law states that the enthalpy level in a phase must be equal to the level of the other except the latent heat produced during the change. This complex phenomenology has been investigated in the past by several authors [1], and various models have been used to face it. It has been shown though, that the size of the mesh grid is a determinant parameter to obtain an accurate solution in the simulation of the phase change problem. Therefore we focus on a simulation of a phase change problem based on cellular automata on variable grid. The cellular automata network approach we have adopted, offers a new methodology; in this approach a complex physical phenomenon is subdivided in components, each model component is represented by a cellular automaton, and the relations between different components are represented by cellular automata interconnections, captured by the cellular automata network abstraction [2]. The cellular automaton *properties* together with their *transition functions* and the interconnections among the automata in the network, are described by way of the *Cellular Automata Network Language* (CANL) [3,4]. CANL is an expression-oriented language, that is cross-compiled into the standard C programming language for sake of performance. In CANL it is also possible to define *global variables*, which represent global features often necessary in describing the modelled physical system.The CANL language is used inside the *Parallel Environment for Cellular Automata Network Simulations* (PECANS) [2].

In building the model of temperature diffusion on a time dependent grid, we analysed three different aspects; first we took into account the temperature variation, according to the Fourier law; based on these results, we provided to change the grid to make it finer in the mushy region; then, the temperature relative to the new grid was calculated by means of an interpolation law. To take into account other physical properties (as, for example, specific heat or conductivity), the user can add other automata to the previous network and he can also change the topology of the network.

The possibility to change the grid size according to the location of the phase change front furnishes a good tool to simulate accurately the jump in the mushy region. Of course the limit of a fine grid size is imposed by numerical stability analysis that imposes a choice of the time step according also to the numerical technique adopted for the simulation; then a compromise between grid size and time step will be imposed in our simulation.

The main results are mainly devoted to display temperature patterns and relative grids during some relevant time steps. The best compromise among mesh grid size and time step is reported.

## 2 Postulation of Heat Diffusion Problem

The problem of unsteady temperature diffusion with liquid/solid phase changing is considered in the present paper. The objective was to create a program which solves the problem of melting of a solid sample, numerically, (i.e. to calculate the position of the solidification front in the liquid-solid system).

The adopted domain is a closed square box inside which a solid sample of sylicium is initially inserted. To both of the end boundaries of the box, cooling or heating are carried out over a particular time interval. The boundary temperatures are, respectively, smaller and greater than the solidification temperature of the liquid. The temperature distribution of the system is calculated via the enthalpy energy equation, that is defined as follows:

$$\rho\, C_p(T)\frac{\partial T}{\partial t} \;=\; \underline{\nabla}\bullet\left(k\underline{\nabla}T\right) \qquad\qquad (1)$$

Of course, the diffusion quantities ($\rho$, $c_p$, k) are supposed to be function of the temperature, and then they assume different values according to control volumes inside which the temperature is higher or lower than the melting one.

The temperature is initially supposed to be uniform along the domain. The boundary conditions are fixed in time and variable in space. We impose different temperatures at the left and right sides of the square domain and zero temperature normal derivatives (adiabatic boundary conditions) at the top and bottom sides.

The region across which the phase change occurs is commonly referred as mushy region. In this region the temperature is discontinuous and a jump between the energy levels of the solid and liquid phase is present. The Stefan law states that this jump is balanced by the so called latent heat of fusion.

Therefore we would stress that one must face two crucial points in the simulation of a phase change problem : i) a good numerical scheme that is fast enough to simulate the diffusion problem without any computational effort; ii) a numerical modelling of the mushy region that accurately calculates the physical quantity transition between the two different phases.

# 3 Time Dependent Grid Problem

The problems described in the previous paragraphs have been analysed in the present paper via the following method. The balance equation (1) has been discretized via a classical explicit finite difference method that allows one to calculate the temperature distribution on two different time levels.

The problem of the simulation of the mushy region has been described elsewhere (see [3]); however, as is well known, there is a strong influence of the step size on this simulation. Some remarks should be made on the choice of the step size. This choice involves a compromise between accuracy and computational effort, as in other numerical methods. The explicit model adopted for our simulation not only suffers the shackles of stability constraints, but also new difficulties that are peculiar to moving boundary problems arise. It is important to decide in advance whether only the freezing rate results are needed, or whether heat flux distributions and heat transfer rates are also needed. As we shall illustrate shortly, the adopted method is capable of accurately predicting freezing rates even with rather coarse grid (5-10 divisions in each coordinate direction).

A rule can be used to find reasonable step sizes, if values are not already prescribed for other reasons: the time step should be small enough that the change in the phase of any node takes many steps in time. In particular, stepping from solid to liquid (or vice versa) in a single step must be avoided completely.

Therefore a good compromise is needed if phase change simulations are requested. The model used in this paper is based on time dependent and non uniform grid, that allow us to calculate accurately the time step in the region where the interface solid-liquid is located. The grid is varied in space, according to a definite law that furnishes the mesh grid via cubic law functions, in the neighbourhood of the node where the solid-liquid phase changing is simulated. Also, the grid is self adaptive in time in the sense that it is able to move itself according to the position of such node. The new nodal values are calculated via a third order interpolation, looking at the new positions of each node and those of the two sandwiching nodes.

# 4 Introduction to the CANL Language

CANL (*Cellular Automata Network Language*) is a high level language specific for programming cellular automata networks; its syntax looks like the Lisp language's one.[3,4]. In the following subsections we describe how to define in CANL the structure of a network of cellular automata, how to define a single automaton of the network, and then the CANL features that take into account the extensions we introduced in our cellular automata model, i. e. the global variables.

## 4.1 Definition of a cellular automata network

A network of cellular automata is represented in CANL by a graph, where each node represents an automaton, and an edge a precedence relation. Each automaton is denoted by a *name*, and its behaviour is described by a set of *properties*, a *transition function* and a *neighbourhood type*; the property of an automaton represents a particular physical quantity to be simulated. Each automaton is the *owner* of its properties; the writing right on a property is guaranteed only to the owner of that property, while the reading right can be granted to all automata.

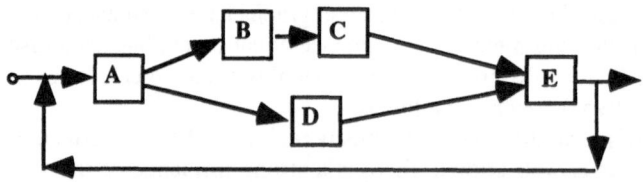

Figure 1: A cellular automata network

Let **A** and **B** be two automata, if one or more properties of the automaton **A** are used inside the transition function of the automaton **B**, then we say that **B** depends on **A** (see Figure 1).

For a network of cellular automata a computational step is obtained through functional composition of all computational steps of each automaton in the network. So, at each computational step, the transition functions of all the automata compounding the network will be executed. According to the precedence relations, a network automaton can be executed only if the executions of the antecedent automata in the network are terminated.

The values of a property for each component correspond to a physical state of the modelled system only at the end of a computational step for the whole network.

## 4.2 Definition of a model component

In CANL each *model component* is described by an automaton. The automaton is denoted by a *name*, and its behaviour is described by a set (possibly empty) of *properties*, by a *transition function* and by a *neighbourhood type*. The properties can correspond to physical quantities (as temperature, mass) or to some features (as the probability of a particle to move). The values of a property can also be real. Anyway, according to the cellular automata model, a property corresponds to a computational grid as in [5].

There are operators that access the *automaton grid* data structure. These operators have been defined as generic operators, so the user can refer to the cell neighbours by simply invoking them on the property's name without having to specify a particular neighbour cell position. For example, the user can refer to the neighbourhood elements of each cell of the property named temperature by invoking (center temperature), (north temperature) or (center_w temperature). The operator in the last expression is used for writing operations, whereas the previous ones are used only for reading operations.

The transition function is built around a set of primitives and user-defined functions, called *auxiliary functions*. For each transition function, the *border conditions* to be used when the transition function is applied to the grid border cells, have to be specified. Currently two types of boundary conditions have been implemented: the *toroidal* one, where the grid is closed on itself, and the *adiabatic* one, where the missing neighbour cell is assumed to be the same of neighbourhood center. For each transition function, *constants* can be used in the transition function body.

The CANL set of primitive functions for model component are: def-automaton, used to define an automaton of the network; def-transition, used to define the automaton transition function; def-aux-fun, used to define the auxiliary functions that may be called inside the transition functions of all the automaton in the network; initialize, used to initialise the grid of a property. See Figure 2 for their syntax.

```
(def-automaton
    :name                         <name>
    :neighbourhood-type           <neighbourhood-type>
    :dimension                    <dimension>
    :border-conditions            <border-conditions>
    :constants                    <constants>
    :transition-function-name     <transition-function-name>
    :list-of-properties           <list-of-properties>)

(def-transition
    :function-name                <function-name>
    :glob-variables               <glob-variables>
    :body                         <body>)

(def-aux-fun
    :function-name                <function-name>
    :type                         <type>
    :neighbourhood-type           <neighbourhood-type>
    :parameters-list              <parameters-list>
    :body                         <body>)
```

Figure 2: Set of CANL primitive functions

The *<body>* of def-function and def-aux-fun is constructed by using the functional composition mechanism and some arithmetic and logical primitive operators, together with the assignment and the conditional operators.

## 4.3 Definition of Global Variables

In making a model of a physical system it is often necessary (and useful) to describe some global features of the system (for example the total mass of a cluster of particles). So we decided to introduce in CANL the possibility to define global features. A global feature is represented in CANL by a *global variable*. A global variable is defined along with a transition function, whose execution will modify its value. It should be noticed that this value can be modified during the execution of the transition function on each cell of the automaton. This value is then available (only in reading mode) to the successive automata in the network.

The definition of global variables is contained in the slot :glob-variables of the primitive operator def-transition.

# 5 CANL Representation of Heat Diffusion on a Time Dependent Grid

## 5.1 Representation of a Variable Grid in CANL

Let us consider a property T on a variable grid. To simplify the description, we have supposed the grid is not uniform only in the X direction.

In this case we still represent the property T by the same matrix of elements, but we also introduce the property X, to take into account the non uniform grid. Each cell of the property X contains the X coordinate relative to the cell in the variable grid. The variable grid and the corresponding properties are shown in Figure 3.

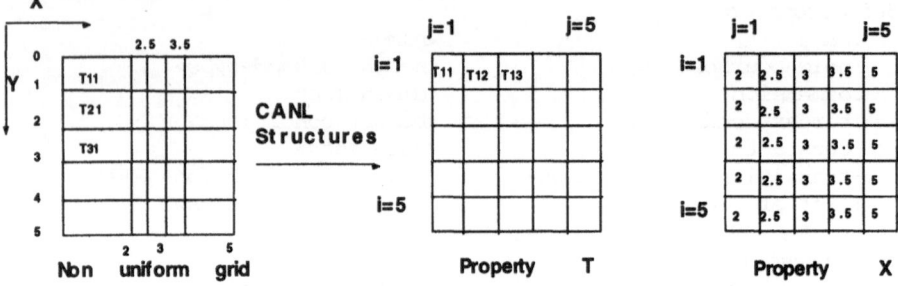

Figure 3: A variable grid and the corresponding properties

## 5.2 Description of network components

We have considered a 2-D model system; the sample is bounded by two lateral plates that have a constant temperature (40°C and 60°C respectively), and by two adiabatic walls. The sample has an initial T=40°C; the melting temperature is Tm=50°C.

Owing to the chosen boundary conditions, the sample will be characterised by isothermal areas that are parallel to the two plates.

In our simulations we are principally interested in examining the behaviour of the sample in the mushy region. To this end we want to consider a finer grid in this region. We have used the CANL language to represent different components of the model by means of a network of cellular automata, as shown in Figure 4.

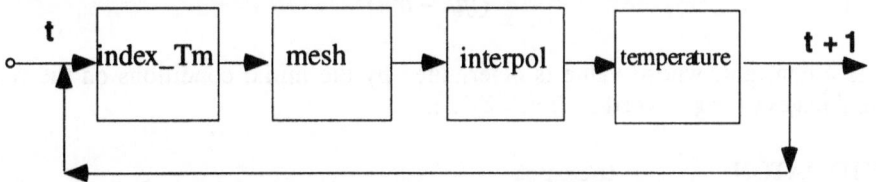

Figure 4: Automata network structure for heat diffusion simulation

The automaton index_Tm provides to determine the mushy region; based on these results, the automaton mesh provides to change the grid to make it finer in the mushy region; then the automaton interpol calculates the temperatures relative to the new grid, by means of an interpolation law; these temperatures are utilised by the automaton temperature, that takes into account the temperature variation, according to the Fourier law.

In the following the four automata compounding the network are described in more detail.

| AUTOMATON | index_Tm |
|---|---|
| PROPERTIES | index_X int |

The property index_X represents for each cell its $j$ index. We need to know this index for each cell, in order to determine the mushy region. In fact, in the case we have analysed there are isothermal areas, and then the $j$ index of the matrix is sufficient to locate the mushy region (the global variables Itm and Xtm). To do it, the transition function of this automaton looks for the cells with the highest temperature lower or equal to the melting temperature.

| AUTOMATON | mesh |
|---|---|
| PROPERTIES | Xt_prec double |
| | X double |

The property x represents the X coordinate relative to the grid cells, whereas the Xt_prec property has been introduced to take memory of the X coordinate relative to the precedent mesh grid, in that the automaton interpol uses both the old and new mesh grid values.

The property Xt_prec is updated at each step with the values of property X. To calculate the new X coordinate, we have chosen to use a cubic function, that is the most suitable for the objective of the present problem. The new X coordinates have been calculated from those at time $t$ as:

$$X_{center} = Xtm + C*(index\_X_{center} - Itm)^3$$

$$C = \begin{cases} index\_X_{center} < Itm & C^- = \dfrac{Xorig - Xtm}{(Iorig - Itm)^3} \\[2ex] index\_X_{center} \geq Itm & C^+ = \dfrac{Xfin - Xtm}{(Ifin - Itm)^3} \end{cases}$$

$C$ is a constant, whose value is determined by the initial conditions on the two lateral plates ($Iorig$, $Xorig$, $Ifin$, $Xfin$).

AUTOMATON   interpol
PROPERTIES   T double

The property T refers to the temperature of the sample.

Once the new X coordinate have been calculated, the new temperatures have been evaluated via a quadratic interpolating function. This function allows one to accurately calculate the new temperature distribution, even if it could not preserve the algebraic monotonicity. Therefore it is possible to define the transition function for the automaton interpol component as:

$$T_{center} =$$

$$T\_new_{center} * \frac{X_{center} - Xt\_prec_{west}}{Xt\_prec_{center} - Xt\_prec_{west}} * \frac{X_{center} - Xt\_prec_{east}}{Xt\_prec_{center} - Xt\_prec_{east}} +$$

$$+ T\_new_{west} * \frac{X_{center} - Xt\_prec_{west}}{Xt\_prec_{west} - Xt\_prec_{center}} * \frac{X_{center} - Xt\_prec_{east}}{Xt\_prec_{west} - Xt\_prec_{east}} +$$

$$+ T\_new_{east} * \frac{X_{center} - Xt\_prec_{center}}{Xt\_prec_{east} - Xt\_prec_{center}} * \frac{X_{center} - Xt\_prec_{west}}{Xt\_prec_{east} - Xt\_prec_{west}}$$

The temperature T is a function of the temperatures T_new calculated by the automaton temperature in the previous time step of the network (T_new is a property of the automaton temperature, that also represents the temperature of the sample), and the X coordinates relative to the old mesh grid (xt_prec) and the X coordinates relative to the new mesh grid (x).

AUTOMATON   temperature
PROPERTIES   T_new double
        recip int

The property T_new represents the temperature of the sample (as the property T of the automaton interpol), whereas the property recip takes account of the two lateral plates at a constant temperature. The temperature variation is calculated according to the Fourier law, starting by the temperatures of the neighbour cells; their contributions have different weight, according to their distance from the center of the neighbourhood:

$$T\_new_{center} = T_{center} + \alpha * [(\frac{T_{east} - T_{center}}{X_{east} - X_{center}} + \frac{T_{west} - T_{center}}{X_{center} - X_{west}}) *$$

$$* \frac{2}{X_{east} - X_{west}} + T_{north} + T_{south} - 2 * T_{center}]$$

being assumed equal to 1 the step size in normal direction. The value of $\alpha$ is limited by stability constraints. Following the Neumann stability method and according to the chosen law for the updated mesh grid, this constrain becomes: $\alpha \leq 1 / 2[1 + 1 / C^2]$. Therefore one could calculate $\alpha$ at each iteration by looking at the minimum between $C^+$ and $C^-$. Nevertheless one could estimate initial value of $\alpha$ by putting $C$ equal to $1/N^2$, being $N$ the grid dimension; after a certain number of iterations one could fix the value of $\alpha$ according to the criteria described above with the calculated value of $C$. The new temperature is a function of the temperatures T calculated by the automaton interpol and the X coordinates relative to the new mesh grid (x).

# 6 Preliminary results and discussion

We solved by means of cellular automata the problem of temperature diffusion with phase change on three different grids. The first case considered is based on a uniform grid, that is the classical grid having equal sizes cells. The second grid considered is a variable grid, that is the grid having different sizes cells. The cells close to the left and right sides are closer to each other with respect to those in the middle of the box. Finally, the last grid considered was a variable and time dependent grid, that is the grid having different size cells at each time iteration. Nevertheless, all the grids have the same number of cells (N=10) that are distributed along the axial direction according to each of the three cases considered.

The law chosen to calculate at each iteration the time dependent grid has been reported in the previous paragraph. The various grids obtained at the beginning of the simulation and after 30000 and 40000 iterations in the case of the time dependent grid are reported in the Figure 5. As expected, the grids become finer according to the location of the melting front.

The temperature distributions (isotherm pattern) obtained after 30000 iterations for each of the three different grids (uniform, variable and time dependent grid) are in agreement with each other confirming that our method based on a time dependent grid is able to calculate with the same number of cells the temperature field that one can obtain with a uniform grid. The great advantage is that the time dependent grid allows one to obtain the same solution with a very fine grid size in the mushy region and coarse elsewhere. This means, in turn, a great reduction of the computational effort, with respect to the case of uniform grid, where the same fine grid size can be obtained only by using very large dimensions.

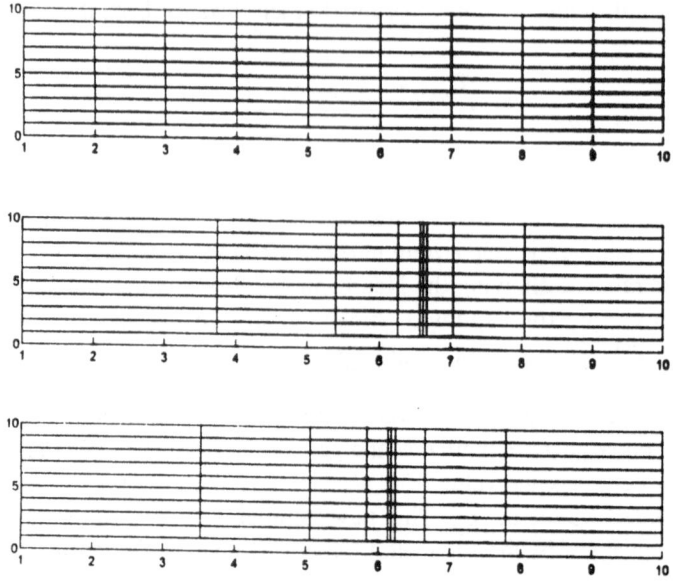

Figure 5: Grid time evolution at t=0, t=30000, t=40000 where t is the number of iterations

We can conclude that our approach based on cellular automata and on a time dependent grid is very promising for future simulations of more complicates problems and different geometries, where it is necessary to calculate accurately the solution only in some regions of the integration domain.

# References

1. V.Voller and M.Cross, "Accurate solutions of moving boundary problems using the enthalpy method.", Int. J. Heat Mass Transfer Vol. 24, 545-556 (1981).
2. M. Mango Furnari, F. Mele, R. Napolitano, "A Parallel Environment for Cellular Automata Network Simulation", in eds. M. Mango Furnari, 2nd Workshop on Massive Parallelism, Capri, Italy Oct. 3-7 (1994) World Scientific Press
3. Carotenuto L., Mango Furnari M., Mele F., Napolitano R., "The Parallel Environment PECANS: a New Methodological Approach for Modelling based on Cellular Automata", in "Atti del Convegno Nazionale Automi Cellulari per la Ricerca e l'Industria, Rende (CS), Italy, 29-30 September 1994".
4. C. Di Napoli, M. Giordano, M. Mango Furnari, F. Mele, R. Napolitano, "CANL: a Language for Cellular Automata Network Modeling", in Proc. of "Parcella 96" - Seventh International Workshop on Parallel Processing by Cellular Automata and Arrays (Berlin, September 16-20, 1996)
5. T. Toffoli and N. Margolus, "Cellular Automata Machines A new Environment for Modelling", MIT Press, (1987).

# CELLULAR NEURAL NETWORKS FOR REALIZING ASSOCIATIVE MEMORIES

Eliano Pessa, Carlo Palma, Maria Pietronilla Penna
ECONA, Interuniversity Center for Research on Cognitive Processing in Natural and Artificial Systems
c/o Dipartimento di Psicologia, Università di Roma "La Sapienza"
Via dei Marsi, 78 - 00185 Roma
Tel.: +39-6-49917631 , Fax: +39-6-4451667 , E-Mail: pessa@axcasp.caspur.it

## Abstract

We studied the performance, as associative memory, of a 1-dimensional Cellular Neural Network, with local connections and a local memorization law of Hebbian type. We found that, contrarily to what expected, this performance was decreasing with increasing neighbourhood amplitude. We proposed some possible explanations of this result, and a speculation about its cognitive meaning.

## 1 Introduction

As it is well known, Cellular Neural Networks (CNN), proposed by Chua and Yang [1], can be considered as a generalization of conventional Cellular Automata, based on the use of analogic units and of neighbourhood functions. The CNN showed very powerful computational abilities in the field of image processing (see, e.g., [2], [3]) and of robot navigation. Less effort, on the other hand, has been devoted to the study of their performance as associative memories. The main problem is that the operation of usual associative memories (as happens, e.g., when we use neural networks) is crucially based on the possibility of doing non-local computations (through long-range connections), whereas the CNN can do, as a principle, only local computations. For this reason it seems impossible, at first sight, to realize through CNN associative memories such as the ones implemented by conventional Hopfield's model or its variants (see, e.g., [4]).

In this paper we present a first study of a possible CNN implementation of an associative memory, based on a suitable 1-dimensional variant of a model proposed by Tan, Hao and Vandewalle [5]. Our analysis deals with the relationship between the "locality" of the computation (measured through the neighbourhood "amplitude" ) and the CNN performance in retrieval phase (measured through the Hamming distance between the retrieved pattern and the stored ones). Very surprisingly, computer simulations up to now performed showed a behavior very different from what expected: CNN performance as associative memory was not a growing function of neighbourhood "amplitude". This seems to imply that local computations could work as well as global ones, and in some cases better. At the end of paper, we will present some mathematical arguments in order to understand the causes of this phenomenon. However, among these latter, there is surely the dimensionality of the model, which prevents us from observing interesting behaviours. Anyway, we point out that, if the best performance of local computations with respect to global ones were confirmed, this circumstance would open the possibility of realizing higher "cognitive" operations by resorting only to local (and whence more economic) computations.

## 2 The model

Our model is realized through a 1-dimensional Time-Discrete CNN (TDCNN). This means that we adopt a discretized time scale $t = 0, 1, 2, ...$ . If we denote by $y_i(t)$ the activation state of the i-th unit at time t , we assume that it will change as a function of the net input arriving to this unit at time t according to a law of the form:

(1) $$y_i(t+1)=F[net_i(t)],$$

where:

(2) $$net_i(t)=S_j w_{ij}\, y_j(t),$$

$w_{ij}$ are the connection weights and F a suitable activation function. As to regards this latter we made the particular choice:

(3) $$F(x)=1 \text{ if } x>=0 \; ; \; F(x)=-1 \text{ if } x<0,$$

i.e. F coincides with the signum function. Relatively to the connection weights we will assume that

(4) $$w_{ij}=0 \text{ when } |i-j| > r,$$

where r is an integer specifying the neighbourhood "amplitude".

The working of our associative memory is based on two phases: the one of memorization, and the one of retrieval. In the former, the connection weights are determined (for a given choice of the neigbourhood amplitude r) as a function of the patterns to be memorized. We assume that the only admissible patterns are the ones constituted by vectors having a number of components equal to the one of CNN units, each component being +1 or -1. The memorization law is nothing but a variant of the so-called Hebbian one:

(5) $$Dw_{ij}^{(s)}=P_i^{(s)}\, P_j^{(s)},$$

where $Dw_{ij}^{(s)}$ is the increment of connection weights due to the memorization of the s-th pattern, whose i-th component is denoted by $P_i^{(s)}$.

At the end of the memorization phase, there is the retrieval phase. Each retrieval process is triggered by the presentation of a retrieval pattern, which coincides whit the initial activation state of CNN units. Then, following a parallel dynamics based on (1), (2), (3), the network state (viewed as a vector whose components are the momentaneous activation states of the network units) can relax towards an equilibrium state, which is to be interpreted as the pattern retrieved from memory. We use the verb "can", because the equilibrium state is not the only attractor possibile for this type of network. To this regard we recall the following

**Theorem**

*A TDCNN, driven by a parallel dynamics according to the rules (1), (2) and (3), if its connection weights are symmetrical, can have only two types of attractors: an equilibrium state or a cycle of length two.*

*Proof.* The memorization rule (5) gives rise obviously only to symmetrical connection weights. Now, let us introduce a Ljapunov function for the network, defined by:

(6) $$E(t) = - S_i |S_j w_{ij} y_j(t)|.$$

Taking into account that:

(7) $$|x| = xF(x),$$

where $F(x)$ is defined by (3), and the activation law (1), it is easy to see that (6) can be rewritten under the form:

(8) $$E(t) = - S_i S_j w_{ij} y_i(t+1) y_j(t).$$

Now, the difference $DE = E(t+1) - E(t)$ becomes, after some transformations:

(9) $$DE = - S_i[y_i(t+2) - y_i(t)][S_j w_{ij} y_j(t+1)],$$

so that, by repeating known reasonings [4], it is possible to see that DE £ 0. The case DE = 0 is realized when:

(10) $$y_i(t+2) = y_i(t)$$

which implies the conclusion. *q.e.d.*

## 3 Computer simulations

In our simulations we used a CNN of N = 680 units, whith non-periodic boundary conditions (zero activation at both ends). We generated 95 vector patterns, each one of 680 elements, whose values +1 or -1 where randomly chosen from a uniform distribution. From this sample we choose a fixed sub-sample of 10 patterns, which were used to realize all memorization phases and to compute the connection weights. As retrieval patterns we used both the memorized patterns and other patterns of the sample not belonging to the sub-sample of these latter. Preliminary experiences showed that cycles of length two never occured, also if they were possible according to the theorem proved in the previous paragraph. We thus decided to stop the network evolution in the retrieval phase at the 250-th time step and to identify the corresponding network state as the equilibrium state, i.e. the retrieved pattern. In any case, we monitored each retrieval process in order to grant for the absence of cycles of length two. This monitoring showed clearly that network evolution in the retrieval phase ended in an equilibrium state much before the 250-th time step.We did numerical experiments by using 5 different values of the neighbourhood amplitude r, ranging from 1 to 5. For each value of r we performed 20 retrieval processes, 10 of which using as retrieval patterns the ones memorized before, and other 10 by using other different patterns, taken from the sample previously described.

As a performance measure of each retrieval process we used the Hamming distance between the retrieval pattern and the retrieved one. The means and standard deviations of these Hamming distances as functions of r are shown below in the Table I, as to regards the memorized patterns, and in the Table II as to regards the other patterns.

| Mean Hamming distance | Standard Deviation | r |
|---|---|---|
| 7.2 | 2.25 | 1 |
| 38.7 | 7.92 | 2 |
| 72.4 | 4.43 | 3 |
| 99.7 | 6.93 | 4 |
| 126.7 | 8.55 | 5 |

Table I

| Mean Hamming distance | Standard Deviation | r |
|---|---|---|
| 9.6 | 2.46 | 1 |
| 38.7 | 4.81 | 2 |
| 78.4 | 7.65 | 3 |
| 102.6 | 6.71 | 4 |
| 130.5 | 8.97 | 5 |

Table II

The data show clearly how the performance becomes worse by increasing r. On the other hand, they put into evidence the smallness of the attraction basins. Practically, starting from $r = 2$ onwards, every memorized pattern cannot be retrieved, and every retrieval dynamical evolution gives rise to a pattern which is far from the retrieval one. We remark also that the mean Hamming distance between retrieval pattern and retrieved one appears to be independent from the fact that the retrieval pattern has been memorized previously or not. It depends only from r. Of course, the performance of this model as an associative memory cannot be considered as good, except the case $r = 1$. In the latter, however, if we use as a retrieval pattern one not memorized before, we obtain as the retrieved pattern more or less the pattern itself. This one, again, cannot be considered as a good performance.

# 4 Mathematical arguments

The observed behavior can be considered somewhat strange. Namely the most known model of associative memory, Hopfield's one, works in an efficient way (more or less), owing to the existence of long-range symmetric connection weights (cfr. [4]). Our model, if we let the neighbourhood amplitude r to tend to the number N of processing units, will become very similar to Hopfield's one (the only difference being that in this latter there are not self-connections). Thus, we could expect that, by increasing the value of r, the performance of our model would become increasingly nearer to the one of Hopfield associative memory. As it is known from the statistical mechanics of this model, its memory capacity is about 0.14 N, which, in our case, being $N = 680$, amounts to 95 patterns (this is the reason for which we choose this number of patterns as the dimension of the sample from which to choose the retrieval patterns). This number is much greater than the number of patterns memorized in our network. We, then, expect that, if r were equal to N, we should have a perfect performance on these patterns (as would happen in the Hopfield model). Besides, with growing r, this performance should improve. In other words, this latter should be worse when $r = 1$, a little better when $r = 2$, and so on. This forecasting is contrary to what observed experimentally.

In order to understand why this happened, we observe that (1), owing to (3), can be written as:

(11)
$$y_i(t+1) = F[S_j S_{(s)} P_i^{(s)} P_j^{(s)} y_j(t)].$$

Now, when the retrieval pattern $y_i(t)$ coincides with one of the memorized patterns, let us say the q-th, we can put (11) into the form:

(12) $$y_i(t+1) = F[P_i^{(q)}C^{(q)(q)} + S_{(s)^l(q)} P_i^{(s)}C^{(s)(q)}]$$

where

(13) $$C^{(a)(b)} = S_j P_j^{(a)} P_j^{(b)}.$$

Owing to the fact that:

(14) $$C^{(q)(q)} = 2r + 1 > 0,$$

we see from (12) that the evolution of the retrieval pattern is determined essentially by the interference terms $C^{(s)(q)}$. Namely the first term inside the square brackets in the right-hand member of (12) has a modulus bounded by $2r + 1$, whereas the modulus of the second term is bounded by $(2r + 1)$ p, where p is the number of memorized patterns. This implies a greater importance of this latter term with respect to the former, if $p > 1$.

By a further rearrangement of the terms in (12), this latter can be rewritten under the form:

(15) $$y_i(t+1) = F[P_i^{(q)}(2r + p) + S_{(s)^l(q)} P_i^{(s)}R^{(s)(q)}]$$

where

(16) $$R^{(s)(q)} = S_{j^l i} P_j^{(s)} P_j^{(q)}.$$

As the range of the sum in (16) depends from r, we expect that, by increasing r, the probability that it will give rise to even greater values will also increase. As a consequence we expect that, by increasing r, the importance of the second term inside the square brackets in the right-hand member of (15) will increase with respect to the first term. This means that, if $P_i^{(q)} = 1$, to make an example, the probability that this second term will have a value lesser than $- (2r + p)$ will grow with growing r. In other words, the probability that, when $P_i^{(q)} = 1$ at time t, the value of $y_i(t+1)$ be -1 will increase with growing r.

This qualitative argument explains why, if we use as a retrieval pattern a memorized pattern, the probability of obtaining an equilibrium (retrieved) pattern different from it, yet at the first time step of retrieval dynamics, will increase with increasing r, as experimentally observed. To this argument, however, can be given also a quantitative form. First of all, we will observe that, for a given value of s, q, i and r, the sum (16) can take all even integer values from 2r to -2r. If we treat each pattern component as a random variable which can assume the values +1 and -1, both with probability 1/2, it is possible to show, by combinatorial arguments, that, for fixed s, q, r and i, the probability that this sum have the value k is given by:

(17) $$p(k) = B(2r, r - (k/2))/2^{2r}$$

where $B(n,m)$ denotes the usual binomial coefficient:

(18) $$B(n,m) = n!/[m! (n-m)!].$$

Now, when we turn to (15), we see that, in the case of the memorization of 10 patterns, we must evaluate the probability that the weighted sum of 9 of these sums will give rise to a value lesser than $-(2r + p)$. To this regard, we assumed implicitely, that $P_i^{(q)} = 1$. However, owing to the symmetry of the probability distributions, the probability would be the same if $P_i^{(q)}$ where -1 and we would obtain a value of this sum greater than $2r + p$. It is easy to see that, being 1/2 the probability of $P_i^{(s)}$, with varying s, of having values +1 or -1, the previous problem reduces to that of evaluating the probability that a sum of 9 variables, each one with the probability distribution (17), will give rise to a value lesser than -(2r+p). Known theorems of probability theory (see, e.g., [7]) show that this probability is given by a convolution product of the form:

(19) $\qquad P(S<-(2r+p)) = S_{-2r(p-1)} £X<-(2r+p) \, P(X)$

where p=10, the number of memorized patterns, and:

(20) $\qquad P(X) = S_{k1,...,k8} \, A(k1)...A(k8) \, G(X,K).$

Besides

(21.a) $\qquad K = k1 +...+ k8$

(21.b) $\qquad A(ki) = B(2r, r - (ki/2))/ 2^{2r}$

(21.c) $\qquad G(X,K) = B(2r, r - ((X - K)/2))/ 2^{2r}.$

The values of (19) can be computed numerically. We obtained

$$P(S<-(2r+p)) = 6,561x10^{-4} \qquad \text{for } r = 1$$

$$P(S<-(2r+p)) = 6,670x10^{-3} \qquad \text{for } r = 2$$

$$P(S<-(2r+p)) = 9,911x10^{-3} \qquad \text{for } r = 3$$

The computation of these probabilities for higher values of r becomes very time consuming, and we don't present here further results. In any case, we see that their values grow with growing r, as expected. Of course, this argument works only for a one-step retrieval dynamics. The issue of the determination of the attractors of this type of associative memory is still open.


# 5 Cognitive implications

The results so far obtained show the existence of precise limitations as to regards the use of CNN for modeling cognitive operation of human or animal memory. Namely, the presence of only local connections seems to be related to a very small attraction basin of each retrieved pattern. This property can give rise to a very low performance. This difficulty cannot be solved by introducing long-range connections because, as it well known from the study of Hopfield's model, the increase of the radius (and whence of the stability) of the attraction basins has as a counterpart a very small memory capacity. In this latter case, therefore, we are faced with two problems: on one hand, the small capacity is not realistic (at least when we want to model long-term human memory) and, on the other hand, we must resort to external criteria for evaluating the model performance. After all, when we retrieve a particular pattern, also if it is characterized by a wide attraction basin, this fact does not grant *per se* that the retrieved pattern is the one we wanted to retrieve!

As we can see, an optimal performance would be obtained if we could use a model taking the advantages of both CNN and Hopfield's network, and rejecting their shortcomings. According to our point of view, a model of this type could be realized through a coupling of a CNN with a second network, designed in order to evaluate CNN's performance in retrieval phase. The parameter r of the CNN should vary as a function of the state of this second network. In this way we could obtain a control of CNN activity, in such a way as to obviate to its errors. Such an architecture is currently under study.

We remember, to this regard, that, when neural-like units are used within models of cognitive processes, such as the ones dealing with human memory operation, these units are mostly viewed as representing cognitive units. This term denotes elementary components of a cognitive structure, such as, for instance, particular features of complex patterns. Then, cognitive processing can be, according to the connectionist point of view [8], considered as the macroscopic result of a cooperative interaction between the signle cognitive units. This is possible, however, only if some form of communication between cognitive units is allowed. And this communication, of course, can be costly for the cognitive system itself, as it requires time, space, and energy. The use of CNN with only local connections appears as a possible framework for doing cognitive computations under reasonable requirements of time and space occupation.

This approach, however, cannot be fully developed owing to two main difficulties:
1) the performance of the models so far realized is very poor;
2) human cognitive system appears to do also nonlocal computations (as happens, e.g., when we make an association between two patterns belonging to domains whose mutual semantical distance is very high, or when we make inferences or deductions which connect different abstraction levels).

Thus, also if the main support to the communication processes between cognitive units seems to be given by local computations, such as the ones done in a CNN, we need also some long range interactions to obtain macroscopic cognitive behaviours. The latter interactions could derive from some collective processes taking place inside systems such as a CNN or Cellular Automata, but could be represented also in an explicit way through models based on a second higher-level neural network controlling a CNN.

## 6 Conclusions

In this paper we presented a model of an associative memory realized through a 1-dimensional CNN with local connections characterized by a neighbourhood amplitude r. We showed, both through mathematical arguments and computer simulations, that the model performance becomes worse by increasing r. We conclude that a low value of r grants for a not so low memory capacity, associated to very small attraction basins. As this latter property appears to be, in some cases, undesirable, it calls for a modification of our original model in order to realize a more efficient associative memory through a CNN.

## References

1. Chua LO, Yang L. Cellular neural networks: Theory. IEEE Transactions on Circuits and Systems. 1988; 35: 1257-1272

2. Nossek JA: design and learning with cellular neural networks. In: Proceedings of the Third IEEE international workshop on cellular neural networks and their applications (CNNA-94). IEEE, Rome, 1994, pp 137-146

3. Harrer H, Venetianer PL , Nossek JA, Roska T, Chua LO. Some examples of preprocessing analog images with discrete-time cellular networks. In: Proceedings of the Third IEEE International Workshop on Cellular Neural Networks and Their Applications (CNNA-94). IEEE, Rome, pp 201-206

4. Amit DJ. Modeling brain function. The world of attractor neural networks. Cambridge University Press, Cambridge, 1989

5. Tan S, Hao J, Vandewalle J. Cellular neural networks as a model of associative memory. In: Proceedings of the First IEEE international workshop on cellular neural networks and their applications  (CNNA-90). IEEE, Budapest, 1992, pp 26-35

6. Goles E, Fogelman-Soulie F, Pellegrin D. Decreasing energy function as a tool for studying threshold networks. Discrete Applied Mathematics. 1985; 12: 261-277

7. Ventsel H. Théorie des probabilités. Mir, Moscou, 1973

8. Quinlan P. Connectionism and psychology. Harvester Wheatsheaf, New York, 1991

# A Cellular Automata Approach to the Simulation of a Prey-Predator System of Mites

S.Bandini and R.Casati
Department of Computer Science, University of Milan
via Comelico 39, I-20135
Milan, Italy
E-mail: bandini@dsi.unimi.it

## Abstract

The classical modeling approach to prey-predator systems relies on extensions of Lotka-Volterra differential equations.

In this paper a Cellular Automata-based model is proposed as an alternative approach for the simulation of a two population ecosystem of mites.

It is shown how several complex features affecting mite population evolution, such as the egg disclosure time, the sexual maturation time, the limited life time, the limited survival capability of predators in fasting condition, and juvenile mortality, can be embedded in the Cellular Automata framework. Preliminary simulation results, together with a comparison between data of a real experiment and data from simulation experiments, are reported. Experimental data are fitted quite well by predictions of the Cellular Automata-based model, whereas they are not by results from the analytical Lotka-Volterra model.

## 1 Introduction

In recent years we have witnessed a widespread interest in ecosystems, both from a theoretical, and from a practical point of view. Most studies have been devoted to modeling ecosystems, in order to devise tools enabling predictions about their evolution. This interest has grown also as a consequence of the introduction of biological control techniques - seen as an alternative to the use of chemical substances - in the treatment of pest infestations. In this case, the goal is to stabilize pest population at a low and controlled level by the introduction of other biological species (predators) that feed with the harmful ones (preys), avoiding at the same time the extinction of all the species involved [1]. If predators can feed only on preys, it may happen that they eat all the preys and then starve to death. In this case, the net result is that a lot of money are spent for buying predators, and, after a while, the plantation is still exposed to a new parasite infestation. On the whole, if this happens, there is no difference with respect to using chemical substances. One would like instead to end up with an ideal situation in which predators keep the number of preys below a given threshold (below a number that the plant can tolerate), predators do not eat all the preys, and survive in a small amount, so that, if a new infestation of preys takes place, there is a maniple of predators ready to neutralize them. In practice one knows (or should know) all the biological parameters which characterize the prey-predator system, and would like to know how many predators to add to the system, in order to reach the ideal situation.

Since it is well known that in certain prey-predator systems the two population oscillate in time, finding the solution of this problem does not seem an hopeless endeavor. Stated in more rigorous terms, one would like to know for which initial conditions the two populations oscillate, the average number of preys is kept below a given threshold, and the whole system is stable, that is an external perturbation (unfortunately not small in general) such as an abrupt increase in the number of preys is automatically neutralized. If this "dynamical equilibrium condition" is reached, a lot of money can be saved (predators are bought just once, in the right amount, and that's all) and the plantation is protected against further infestations.

When dealing with ecosystems constituted by populations of interacting biological beings, in which the interaction is of the prey-predator type, the classical approach has been the use of Lotka-Volterra (L-V) equations [2, 3, 4] with their extensions [5, 6, 7, 8], and their discretized versions in which they are seen as dynamical systems [9, 10].

Recently the simulation of prey-predator systems has been addressed also by an alternative approach which is based on Cellular Automata (CA) models [11, 12, 13, 14, 15]. These models assume that individuals of both populations are distributed on a lattice, and that interactions among them take place over a predefined neighborhood. The CA evolution rules are defined in order to reproduce the elementary biological processes that rule the dynamics of the ecosystem. The characteristic curves of population evolution over time result from the application of the simple local CA rules all over the lattice.

In this paper a very simple CA model for the simulation of a prey-predator system will be presented. Moreover, it will be shown how this model can be enhanced to tackle the simulation of a specific two population system of mites (*Tetranychus urticae* Koch as prey, and *Amblyseius californicus* McGragor as predator). This particular ecosystem is being studied at the Istituto Sperimentale per la Zoologia Agraria, near Florence, with the objective of controlling infestations in vineries (glasshouses for vines).

In this perspective we claim that the CA approach will lead in a near future to the possibility to simulate the real ecosystem, enabling to test and to optimize pest control strategies through computer experiments.

In the following section we present a brief survey of the classical approach with some of its shortcomings. In Section 3 we define the basic prey-predator CA model and present some simulation results showing oscillatory behavior in the evolution of the two populations. The enhancement of the model for the considered mite ecosystem is presented in Section 4. Preliminary simulation results of the enhanced model, together with a comparison between data from simulation experiments and data from a real experiment monitoring population growth, are reported in Section 5. Conclusions are drawn in Section 6.

# 2 Survey of the Classical Modeling Approach to Prey-Predator Systems

The classical approach to the study of the evolution of a prey-predator system is to write down a system of differential equations for the two populations. Generally the starting point is the system of L-V equations [2, 3, 4].

$$\begin{cases} \dfrac{dx}{dt} = \alpha\, x - \beta\, x\, y \\[2mm] \dfrac{dy}{dt} = -\,\gamma\, y + \delta\, x\, y \end{cases}$$

x and y are time-dependent functions that represent respectively the number of preys and the number of predators; $\alpha$, $\beta$, $\chi$ and $\delta$ are fixed positive constants. The 4 terms in the above equations have the following meaning:

- in absence of predators (y=0), the number of preys increases in time proportionally to the number of existing individuals ($\alpha$ is the growth rate); this implies that in absence of predators, the number of preys increases exponentially (malthusian law of growth);
- in absence of preys (x=0), the number of predators decreases in time proportionally to the number of existing individuals; this implies that in absence of preys, predators decrease exponentially;
- in presence of predators, the variation in time of the number of preys is affected by a decrease due to encounters with predators, and this effect is assumed to be proportional to the product of both populations;
- in presence of preys, the variation in time of the number of predators is affected by an increase due to the welfare provided by feeding on preys; this effect is assumed to be proportional to the product of both populations.

Solutions of the above system of equations can be found in current literature. See for example [16] and references in [1]. L-V equations are characterized by two stationary points that are (0,0) (saddle) and $(\chi/\delta, \alpha/\beta)$ (center). It can be shown that oscillation of the two populations is a feature that holds for every initial condition different from the two stationary points. For example, the solutions of the L-V system with the following parameter $\alpha = 0.2$, $\beta = 0.01$, $\chi = 1$ and $\delta = 0.1$, and for the following initial condition x(0) = 10, y(0) = 10 are plotted in Figure 1.

138

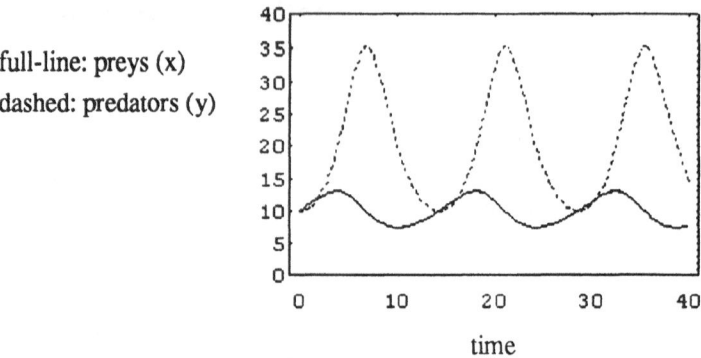

full-line: preys (x)

dashed: predators (y)

time

Figure 1: Oscillating behavior of the solutions of L-V equations

By adding other terms, L-V equations can be extended to take into account other processes, such as external removal of individuals, competition among individuals of the same species (e.g. due to food shortage), growth rate saturation, and so on [5, 6, 8]. For example, in order to take into account the fact that the environment can only support a maximum number X of preys, it is possible to hypothesize that the growth rate

$$\alpha = \frac{x(t + dt) - x(t)}{x(t)\, dt}$$

is not constant but proportional to the difference between X and the current number of preys.

$$\alpha(t) = a\,(X - x(t))$$

This is the so called Verhulst's growth saturation. Or, just to make another example, to take into account spatial inhomogeneities and motion of individuals, a diffusive term can be incorporated into L-V equations [7, 13].

It is possible to go on and on, adding new terms to the population equations, by which other processes are modeled. This is what we intend for "Classical Modeling Approach".

However, despite L-V equations and their refinements have proved to be very useful in modeling a large number of biosystems, the classical approach is affected by at least two shortcomings.

- Although the equations can be generalized to describe spatial inhomogeneities and the individuals' motion by adding a diffusive term, in general, it is not easy to investigate the time dependent spatial distribution of individuals [11]. In fact, many mathematical models have shown that adding a spatial dimension and dispersal to population processes alters population dynamics [17]. Moreover, it can be shown that with a diffusive term the short range character of the interaction between preys and predators cannot be taken into account correctly [13].
- Using either differential equations, or their discretized versions in which they are seen as dynamical systems[9, 10], it is prohibitively complicate to introduce

into the model complex features (such as for example photoperiodism, i.e. the dependence of egg and sexual maturation time on the ratio of day/night lengths) which may turn out to be fundamental in certain ecosystems. If the complexity of the considered mechanism is very high, it may turn out that writing down an equation is impossible.

Since the specific two population system of mites we have considered is characterized by complex features such as the above mentioned, we chose to model the system using the CA approach.

# 3 A Basic CA Model for Prey-Predator Systems

The simplest CA model for the simulation of two interacting populations embeds the same processes taken into account by L-V equations. The model we present here is the direct formalization in the language of CA of the "game of transformation" described by Eigen in [18].

The individuals of the two populations are assumed to live and interact on a two-dimensional square lattice. Each site on the lattice can host only a single individual which is described by means of a discrete variable of state. Different values of the variable of state correspond to different entities (prey or predator).

Following the formal definition of CA [19], according to which a Cellular Automaton is a 3-ple $A = \langle G, Q, f \rangle$, where G is the two-dimensional lattice, Q is the finite set of the values of the variable of state, and $f : Q^{|L_0|} \times Q \to Q$, with $L_0$ being the set of the neighborhood sites, is the next state function, in our case we have:

- $G = \{(i, j) : i, j = 1, 32\}$;
- $Q = \{0, 1, 2\}$ where 0 represents the empty site, 1 denotes the presence of a prey, and 2 denotes a predator;
- the neighborhood $L_0$ is of the Moore type, thus $|L_0| = 8$;
- the next state function is comprised of 3 updating rules which model the processes of birth, death and reproduction for both populations. Let *Cond1* be the minimum number of preys present in the neighborhood of an empty site necessary to give birth to a prey in the empty site. Let *Cond2* be minimum number of predators in the neighborhood of a prey necessary to substitute the prey with a predator, and $n_{i,j}(k)$ (k = 0,1,2) the number of sites in the neighborhood of (i,j) which are in the k state. Let $x_{i,j} = \{x(i,j), x(i+1,j), x(i-1,j), x(i,j+1), x(i+1,j+1), x(i-1,j+1), x(i,j-1), x(i+1,j-1), x(i-1,j-1)\}$ be the vector of dimension 9 which belongs to the next state function f domain. In the following we will denote with $x_{i,j}^{\alpha}$ a vector of the function f domain whose first component $x(i,j) = \alpha \in Q$ is fixed, whereas all other components can assume any of the possible values in Q. The next state function $f : Q^{|L_0|} \times Q \to Q$ is the following:

$$f(x_{i,j}^{0}) = \begin{cases} 1 \text{ if } n_{i,j}(1) \geq \textit{Cond1} \\ 0 \text{ otherwise} \end{cases}$$

$$f(x_{i,j}^{1}) = \begin{cases} 2 \text{ if } n_{i,j}(2) \geq \textit{Cond2} \\ 1 \text{ otherwise} \end{cases}$$

$$f(x_{i,j}^{2}) = \begin{cases} 0 \text{ if } n_{i,j}(1) = 0 \\ 2 \text{ otherwise} \end{cases}$$

The above updating rules can be expressed in natural language as follows:

- if an empty site has at least *Cond1* preys in the neighborhood, then (at the next time step) a prey is generated in the empty site;
- if a prey has at least *Cond2* predators in the neighborhood, then (at the next time step) the prey becomes a predator; otherwise the prey remains unaltered;
- if a predator has no prey in the neighborhood, then (at next time step) it dies.

For some values of *Cond1* and *Cond2* and some initial conditions the two populations oscillate. For example, simulation results obtained with the above basic model, with an initial number of 100 preys and predators respectively, and *Cond1* = 1 *Cond2* = 1 are plotted in Figure 2. The two populations evolve toward a stable oscillatory regime after an initial transient. In Figure 3 we show the phase space with the appearance of a closed orbit.

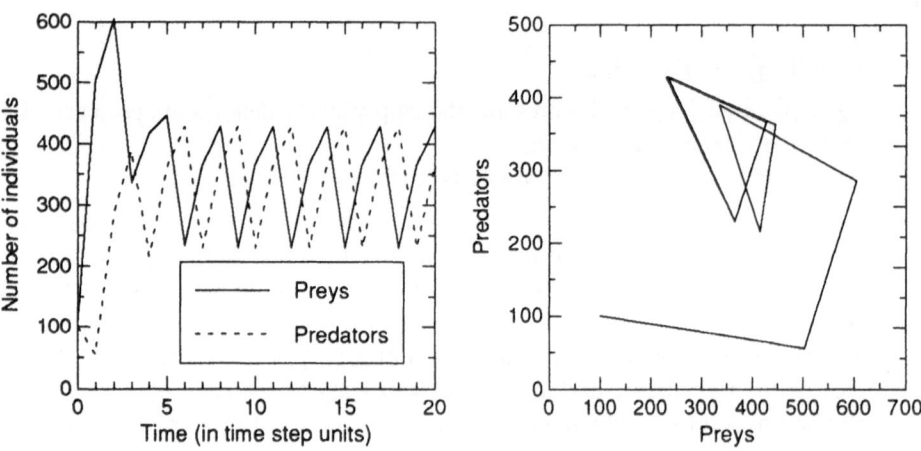

Figure 2: Evolution in time of the populations    Figure 3: Trajectory in the phase space

To obtain from the basic model the oscillatory behavior of the two populations was a fundamental checkpoint for the model, in order to show the feasibility of the CA approach to the simulation of this kind of systems.

However CA permit, even in this simple case, a much more detailed understanding of the dynamics, as they enable to monitor in time also the evolution of the spatial distribution of the two populations. In Figure 4 we report a sequence of snapshots of the "leaf" at different time steps of the simulation. Dark, light gray and white represent preys, predators and free leaf respectively.

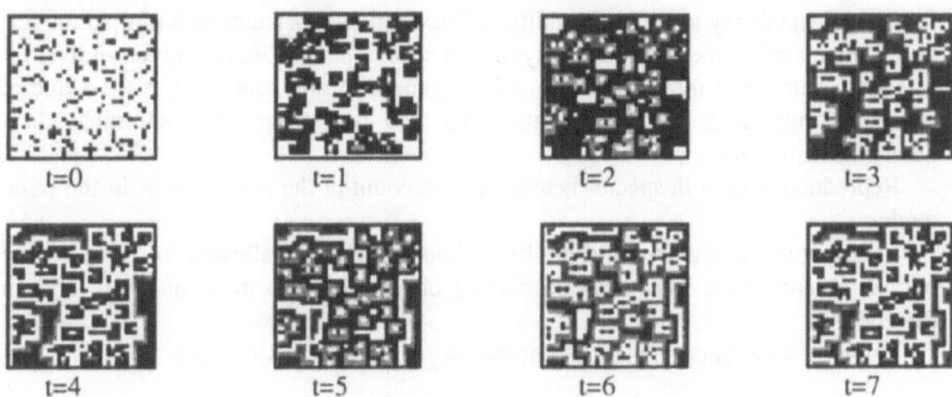

Figure 4: Snapshots of the "leaf" at the first time steps of the simulation

It is clear that this approach allows to follow interesting effects which cannot be described so easily by the classical one. For example, it is possible to follow directly the process of clustering, and to compute the distribution of cluster sizes.

# 4 The Enhanced Cellular Automata Model for Mites

The study of the real prey-predator ecosystem made up of two specific species of mites (phytofagous *Tetranychus urticae* Koch as prey, and phytoseeids *Amblyseius californicus* McGragor as predator) served as a stimulus to enhance the basic model by introducing a set of features which characterize prey-predator systems in general and the two considered populations of mites in particular.

In all prey-predator systems like the one under study in which both of the two species is oviparous, it is fundamental to take into account the egg disclosure time and the sexual maturation time. Hence, in the enhanced model the prey population is split into eggs, immatures and adults. Immatures differ from adults as they are not capable of reproduction. The predator population is subdivided into eggs and adults only. Eggs of both species are characterized by a disclosure time, i.e. they disclose giving rise to an immature individual in the case of preys, or to an adult in the case of predators, after this time has elapsed since deposition. Disclosure times for prey eggs and for predator eggs can be different. Immature preys are characterized by a sexual maturation time, i.e. after that time has elapsed they transform into adults being capable of reproduction.

Adults (both preys and predators) have a limited lifetime, i.e. after that this time has elapsed they die. In the present model this time is assumed to be equal for preys and predators.

Predators have also a survival time which is the time they can survive without eating. A time-counter is associated to each predator representing the energetic reservoir of the mite. Each time step it is decreased by one unit. When it is zero the mite dies, while it is reinitialized to its maximum value when the predator eats a prey. All these times are assigned in timestep units, which we assume to correspond to 24 hours.

The effect of prey juvenile mortality is described by a percentage input parameter, and is simulated in the following way: at the second day of life of a given individual a random number in the interval [0, 100] is generated and compared to the value of the percentage. If it is less or equal than the percentage, the individual dies, otherwise it survives.

Reproduction of both species is taken into account in the same way as in the basic model.

Finally, prey immatures and adults of both species are allowed to move on the lattice. At each time step they are shifted one lattice step in a randomly chosen direction.

The next state function is thus conceptually divided into two steps:

- updating of the internal state of the individual that occupies the site (eggs and immature mature, adults get older, predators that are not eating consume part of their energetic reservoir, etc.); and substitutions that take into account the birth and the death of preys and predators;
- motion.

As the next state function consists of one part that is applied synchronously (the first), and one part that is applied sequentially that models the motion of individuals, this is a site-exchange CA [13].

Every simulation starts by assigning to each individual a random position on the lattice, and a random age.

We notice that this model is very similar to the WATOR game by Dewdney [20], even if there are some slight differences in the features that we have considered.

From the computational point of view, in the enhanced model individuals are described as agents, and are implemented as objects by means of an object-oriented programming language.

# 5 Preliminary Simulation Results of the Enhanced Model and Comparison with Experimental Data

As we have already mentioned, in this specific mite prey-predator system taking into account the process of egg maturation is fundamental in order to correctly simulate the dynamics of the system. In particular, for the two considered species the egg disclosure time is a biological parameter that depends on the ratio of day/night lengths. As a first step toward considering this complex dependence, and with the aim of testing the sensitivity of the model to this parameter, we performed two simulations setting different values for "Prey egg opening time".

| Initial number of preys | 100 |
|---|---|
| Initial number of predators | 100 |
| Prey egg opening time | 2 or 3 |
| Predator egg opening time | 0 |
| Prey sexual maturation time | 0 |
| Life time | 8 |
| Predator survival time | 1 |
| Prey juvenile mortality | 0 |
| Cond1 | 1 |
| Cond2 | 1 |
| Motion | Yes |

Table 1: Input of the two simulations

We report in Figure 5 and Figure 6 the corresponding plots.

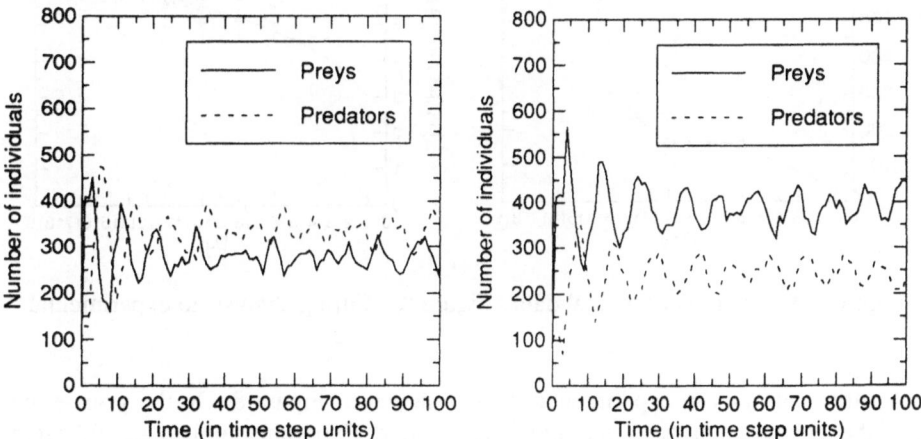

Figure 5: Prey egg maturation time = 2    Figure 6: Prey egg maturation time = 3

The number of individuals on the y axis for preys is the sum of eggs and adults. It is possible to see how the change of such a parameter affects dramatically the evolution of the two populations. With maturation time equal to 2, the number of preys is smaller than the number of predators, while in the other case predators are always fewer than preys. Moreover, recalling the real problem of keeping under control the number of preys (which is the plant-harmful species), the comparison of these results shows that with maturation time equal to 2, an initial number of 100 predators suffices to keep the number of preys at an average of 300, while if the maturation time is 3, with 100 predators the average number of preys is greater (about 400). The difference of 100 preys on the average may turn out to be fundamental for the fate of the leaf.

These results confirm the criticality of this parameter. Moreover they show how CA simulations may be used to make predictions about the evolution of the ecosystem, providing a tool to test biological pest control strategies in complex ecosystems.

A real experiment monitoring the evolution of the two populations was performed at the Istituto Sperimentale per la Zoologia Agraria near Florence [21]. In particular, the growth of preys alone was followed on 12 distinct leaves of approximately the same size (independent trials) during 19 days. As initial condition, 10 fecundated female preys were placed on the leaves. After 6 days the total number of prey individuals present on each leaf was determined. Then, every two days during approximately 2 weeks, the number of preys per leaf was recorded. The plot of the average number, over the 12 leaves, of preys as a function of time (Figure 7) shows that preys increase until the 17th day, while at the 19th day a prey population decrease was recorded because of the reaching of the critical density, that is Verhulst's saturation.

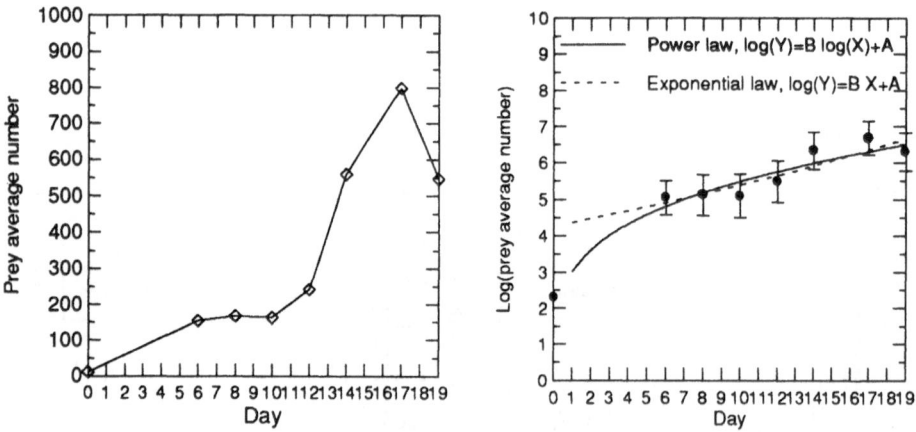

Figure 7: Plots of experimental data   Figure 8:   Fitting analysis to experimental data

A fitting analysis to experimental data is shown in Figure 8. On the y scale the log of the average number of individuals is reported. Experimental data are indicated as dots with the corresponding error bars. The power fit coefficients are approximately B=0.26 and A = 1.1, while for the exponential B = 0.02 and A = 1.45.

In the first part of the plot (from time step 0 to 6) a difference between the power and the exponential fitting curve is clearly evident: the power curve seems to fit the first two experimental data much better than the exponential.

Although this is just a first graphical comparison, strongly demanding further analysis, it suggests that in the considered range the power law describes the evolution trend better than the exponential. This fact is remarkable since, being the exponential the solution provided by L-V equations, it would represent a further indication of the inapplicability of this model to this specific case of study.

It may be argued that comparing data from the real experiment to the exponential is not fair since the leaf is intrinsically saturable (due to its limited size) while the exponential is the solution of the pure L-V equation system, i.e. without Verhulst's growth rate saturation term. However, the largest difference arises in the first 6 time steps when the prey population is still small, the effects of saturation have not come into play yet [21], and it is thus reasonable to neglect the saturation term in the equations.

Then we performed a campaign of simulation experiments with the aim of obtaining the experimental growth curve of preys. The best resembling result was obtained using inputs of Table 2:

| Initial number of preys | 10 |
|---|---|
| Initial number of predators | 0 |
| Prey egg opening time | 0 |
| Predator egg opening time | 0 |
| Prey sexual maturation time | 0 |
| Life time | 1000 |
| Predator survival time | 0 |
| Prey juvenile mortality | 0 |
| Cond1 | 2 |
| Cond2 | 0 |
| Motion | Yes |

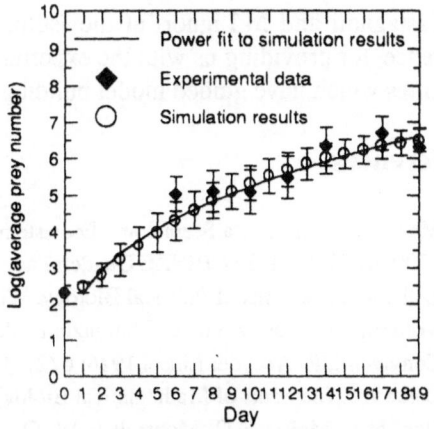

Table 2: Input of simulations

Figure 9: Comparison between simulation and experimental data

As all simulation experiments were shorter than 1000 time steps, setting Life time=1000 means assuming that mites do not die for natural causes (i.e. this is a process that can be neglected). In order to compare simulation results with experimental data, we performed 12 simulation runs, all with the same input parameters of Table 2, but with a different seed for the random number generator which governs mites' motion (simulating in this way the independence of real experiments). The logarithm of the average number of individuals (over the 12 experiments) present on the lattice at each time step as a function of time is reported in Figure 9, together with a power law fit to simulation data.

It is possible to see how simulation data are well fitted by a power law (B = 0.36, A=0.82), which is also the kind of law which best fits experimental data. Moreover, also the extension of the error bars of the experimental data and of the data from simulation experiments is quite similar, providing evidence for the possibility to compare CA predictions with real data.

# 6 Conclusions

We have presented a basic CA model for the simulation of a prey-predator system. It is very easy, but it captures very agilely an essential feature of real world prey-predator systems such as the oscillation in time of the two populations. Moreover, it lends itself to extensions in a very natural way, enabling the representation of the fundamental characteristics of a particular mite ecosystem to be taken into account. In the case of the growth of preys alone, experimental data are fitted quite well by predictions of the proposed CA-based model, whereas they are not by results from the analytical Lotka-Volterra model. This indicates that the CA approach is very promising, even though a lot of work is still necessary before it can be applied with success as a reliable predictive tool in the study of real ecosystems.

## Acknowledgments

The authors wish to thank M.Milani at the University of Milan, and M.Costato at the University of Modena for their leading role in this research, as well as M.Castagnoli and M.Liguori at the Istituto Sperimentale per la Zoologia Agraria, Florence, for providing us with the experimental data, and the biological information on mites which have guided model building.

## References

1.  Maini S, Nicoli G. La Serra come Ecosistema. In: Celli G (ed) Ecosistemi. LE SCIENZE, 1990, pp 37-43 (LE SCIENZE Quaderni no. 53)
2.  Lotka A J. Elements of Physical Biology. Williams and Wilkins, Baltimore, 1925
3.  Volterra V. Variazioni e Fluttuazioni del Numero d'Individui in Specie Animali Conviventi. R. Acc. dei Lincei 1926; 6 (2): 31-113
4.  Lotka A J. Elements of Mathematical Biology. Dover, New York, 1956
5.  Goel N S, Maitra S C, Montroll E W. On the Volterra and other Non-linear Models of Interacting Populations. Reviews of Modern Physics 1971; vol. 43, no. 2: 231-276
6.  May R M. Stability and Complexity in Model Ecosystems. Princeton University Press, Princeton, 1975
7.  Okubo A. Diffusion and Ecological Problems: Mathematical Models. Springer-Verlag, 1980
8.  Ritman E. Genesis Redux: Experiments Creating Artificial Life. Windcrest/McGraw-Hill, 1994
9.  Sandefur J T. Discrete Dynamical Systems. Clarendon Press, Oxford, 1990
10. Jeffries C. Mathematical Modeling in Ecology. Birkhauser
11. Satoh K. Computer Experiment on the Complex Behavior of a Two-Dimensional Cellular Automaton as a Phenomenological Model for an Ecosystem. Journal of the Physical Society of Japan 1989; 58: 3842-3856
12. Ermentrout G B, Edelstein-Keshet L. Cellular Automata Approaches to Biological Modeling. J.Theor. Biol. 1993; 160: 97-133
13. Boccara N. Automata Network Models of Interacting Populations. In: Goles E, Martinez S (eds) Cellular Automata, Dynamical Systems and Neural Networks. Kluwer Academic Publishers, The Netherlands, 1994, pp 23-77
14. Molofsky J. Population Dynamics and Pattern Formation in Theoretical Populations. Ecology 1994; vol.75, no.1: 30-39
15. Sutherland B R, Jacobs A E. Self-Organization and Scaling in a Lattice Predator-Prey Model. Complex Systems 1994; 8: 385-405
16. Rosen R. Dynamical Systems Theory in Biology. Wiley-Interscience, 1970
17. Kareiva P. Population Dynamics in Spatially Complex Environments: Theory and Data. Philosophical Transactions of the Royal Sociaty of London 1990; B 330: 175-190
18. Eigen M, Winkler R. Il Gioco. Adelphi Edizioni, Milano, 1986
19. Goles E, Martinez S. Neural and Automata Networks: Dynamical Behavior and Applications. Kluwer Academic Publishers, Boston, 1990
20. Dewdney A K. The Armchair Universe. W H Freeman, New York, 1988
21. Castagnoli M, Amato F. Studi di Laboratorio sull'Interazione tra il Predatore *Amblyseius Californicus* McGregor (Acarina: Phytoseiidae) e la sua Preda *Tetranychus urticae* Koch (Acarina: Tetranychidae). REDIA 1991; vol. LXXIV, no.1: 77-85

# Urban Cellular Automata: An Evolutionary Prototype

Papini L. & Rabino G. A.

DISET - Faculty of Engineering, Milan Polytechnic

P.zza Leonardo da Vinci 32, 20133 Milan - Italy

tel. 0039-2-2399.4100 / fax 4105, E-mail: rabino@cdc8g5.cdc.polimi.it

## Abstract

The subject of the paper is a cellular automata with evolutionary transition functions, applied in urban context.

Using a cellular automata to describe a phenomenon, the definition of the model consists in building the local transition rules; this process may be supported by a machine learning tool like a Learning Classifiers System.

Because this formalism uses genetic algorithms to select and to find rules, the model is explicitly evolutionary.

## 1 Cellular Automata in Urban Modeling

### 1.1 From global to local description

Urban models belong to a family of models which are extremely delicate to use and interpret since they refer to situations in which the free action of individuals and hence decisions based on attitude and behaviour, are especially important components. Up till now, urban models have essentially been based on the Input-Output Theory, proposed by Ghosh [1958], which expresses in a linear way the interaction between productive sectors, and the Theory of Spatial Interaction [Carrothers 1956] which suggest how interaction parameters between productive sectors vary in function of the spatial location of the single activities.

The two theories are unified and directly applied to an urban context in the Lowry Model [1964], which describes urban activities (housing, commerce, industry) as productive sectors in the Input-Output Theory, interacting in concordance with Spatial Interaction Theory.

By using differential equations to describe these classical models, we are obliged to analyse urban phenomena through spatial averages, in order to satisfy the requirement of continuity and derivability for solving such equations.

Attempts to make global descriptions with differential equations, inevitably run into the difficulty to express phenomena which cannot be described in statistical terms, since they lack in spatial and temporal ergodity.

An alternative approach to global description is to provide a local description or the description of the phenomenon through its parts, reformulating the global model as an aggregation of local descriptions. In urban field the word "local" has an immediate meaning: a whole territory is global and its parts are local.

The Lowry Model already made provision for territorial subdivision in interacting zones in accordance with Spatial Interaction Theory, but the modelling approach is global because the model is a single block of differential equations.

The global description of urban territory is substituted with a local description where every part interacts by means of its own laws with a limited number of neighbouring zones; these zones belong to the "locality" of the zone; the model is made up of many blocks, each with its own dynamic.

This kind of modelling is spontaneously represented by means of a Cellular Automata, whose cells are parts of the territory.

## 1.2 The first conceptual applications

The application of cellular automata in the urban context was made by Tobler [1979], but his work was not continued because of the strength of classical Lowryian models. Couclelis [1985] recently reproposed the use of this formalism, this time the proposal being welcomed and developed by Phipps [1989] and by Cecchini and Viola [1992, 1995] whose works aimed to analyse the behaviour of generic cellular automata, explaining the generic results as representing specific urban and geographical phenomena. Recent developments by Portugali and Benenson [1995], relate to the analysis of the self-organising behaviour of a cellular automata. In their model they emphasise this aspect by studying how population flows affect on house values in residential urban areas. Another important general contribution, applying cellular automata in urban analysis, is the work by Batty and Xie [1994, 1995]. They used stochastic cellular automata, whose transition functions are defined in probabilistic terms, representing a generic territory on which generic two state cells, urbanised or vacant, are localised. The main results are the confirmation of the utility of applying cellular automata in urban modelling, since they help us to identify the emergent evolution of town, based on local interaction.

## 1.3 An example of an operating instrument

The models described above are very simple from the urban point of view, because they make a pure diffusion simulation and because they do not take account of different urban sectors. An attempt to take into consideration the presence of many different urban sectors in modelling with cellular automata and associating many different states to every cell, was made by White, Engelen and Uljee [1993], simulating the evolution of Cincinnati.

The cellular automata is a deterministic automata, whose local transition functions are affected by noise. Every automata module is a 250x250 m. portion of territory. A main land use is assigned to each module and every cell interacts with a constant number of neighbouring cells within a radius of 6 cells (1500 m.). Transition function assigns the next state to the cell, choosing the state which maximises a quantity obtained from the weighted sum of the states of neighbouring cells, each weighted with a parameter which expresses the action a neighbouring cell will have on the different states of a given cell. The states, each cell can get, are divided in two groups: variable an fixed: a fixed cell never changes its state and it corresponds to an exogenous land use such as road, railway, park etc.; the variable states are the endogenous land use such as housing, industry, commerce and unused area.

# 2 Proposal of an evolutionary cellular automata applied to territory

## 2.1 Evolutionary modeling

The description of a phenomenon, possibly a urban one, by means of a cellular automata, apart from the a priori modelling choice lied to the selection of this descriptive tool, model formulation essentially involves the definition of interaction rules.

This definition is a low level process, consisting of building tables containing a correspondent state transition for every possible neighbour and state of a cell. If the neighbour is large the number of rules to definite can be quite high (for k possible state and a neighbour of n cells, the number of rules is $k^n$ or $k^{n+1}$, if the transition rules depend on the state of the cell). In this case, one usually resorts to simplified rules involving a sum or other processes, hence avoiding possible redundancy, but loosing information and reducing the modelling opportunity. Later we demonstrate the utility of supporting this process by the definition of rules through an interface which translates the model-maker's ideas into low level rules. Such an interface takes on task of the translating model axioms into laws, making up the interaction rules, and developing the algorithms to perform the simulation task to the cellular automata. With this formulation, the interface becomes a modelling support instrument. The model-maker has a high level role: he doesn't take part directly in defining the model by writing the low level rules but drives the machine to the right solution. The model is not static and its variation is not exclusively tied to the adaptation of possible parameters, it is dynamic and varies through the model-maker driving, exploiting the machine to explore the possible alternative model formulation.

## 2.2 An evolutionary cellular automata applied to territorial analysis

Looking at previous assertion, we refer to White et al. cellular automata, revolutionising the structure of such model in evolutionary way, referring to the analysis and modelling techniques provided by the modern approaches of cognitive sciences and artificial intelligence, to propose a new approach to urban modelling.

The evolutionary approach is the capacity of the model to modify itself in function of the results of the descriptions it provides, not only by means of the parameter variation, but explicitly through the variation of the way the model is built. The evolutionary capacity of the model is obtained by defining appropriate dynamics to vary the model structure and defining a selection process to determine which models are more appropriate.

The model tries to identify the emergent dynamics from a series of learning maps, representing the development of the territory at different times, aided by the external support of the model-maker.

The machine consists of an asynchronous cellular automata representing the map of the territory whose developing dynamics we wish to identify. This automata is made up of cells representing homogeneous portions of territory, as in the White et al. model, with the difference that the rules are not defined a priori but are defined by means of an evolutionary process.

The system has to learn a series of disjunctive concepts representing the possible global transition of the automata, described with sets of local interaction rules. These rules may be simply structured like Stimulus-Reply (SR) dynamics, where the stimulus comes from the environmental information of each cell, that is the neighbouring state with which the cell interacts, and the reply is the state variation suffered by the cell.

With this formulation, the problem of the identification of interaction rules moves from a classical problem of numerical parameter identification (the interaction parameter of the White et al. model), to a machine learning problem.

If the instrument, used for learning, has adequate feedback, we find a typical problem of learning incremental disjunctive concepts, formed by the reply to the stimulus, driven by opportune heuristic functions called payoff or fitness functions.

Among the approaches available in machine learning, the one most able to identify the model dynamics is a Learning Classifiers System because:

- it allows a simple rules management, each rule being codified by binary strings
- its operation is independent of the environment on which it works but through opportune input-output interfaces.
- the evaluation tool, giving payoff to drive learning, is independent and may be simply interfaced with the cellular automata by means of suitable evaluation methods
- the search methods for possible interaction rules is based on Genetic Algorithms, so that the model is explicitly evolutionary.

## 2.3 Structure of a specific learning classifier system to build urban cellular automata

The system that can integrate a classifiers system with an urban model, is a system operating on an artificial environment that is a cellular automata, representing the territory whose dynamics we have to identify.

### 2.3.1 The rule manager and specific rules

The structure of the rule manager is the same as a standard classifiers system: there are the message list and the rule base. The rules, represented by classifiers, are local transition rules for each automata making up the cellular automata. These are strings, defined on the alphabet {0,1,#} of fixed length, divided in two substrings: the condition part and the action part. The condition part is subdivided in turn, into three sub-strings, representing three predicates: the context predicate, the neighbour predicate and the state predicate. The action part is simply the new state that the cell will obtain. The selection of the classifier candidate to propose an action, is done by unifying messages coming from the environment with the condition part.

Let us now look specifically at the meaning of the different condition parts:

- The zone condition or context predicate is useful for diversifying the behaviour of automata representing different parts of the territory. From the urban point of view, the presence of a zone condition may be justified, for example, by different a interaction process between commercial cells in the centre of a town and in the outskirts.
- The state condition or state predicate identifies the actual state of the cell at each temporal step, activating different behaviour for cells in different state.
- The neighbouring condition or predicate is the fundamental component that allows the selection of the right rule to respond to the environmental stimulus, constituting the environmental state of the finite neighbour with which the cell interacts.

The rules are of a SR type, which simplifies the management of reward and penalty assignment.

### 2.3.2 The interaction interface

The interaction interface is made up of the input output modules. The input module is concerned with codifying the zone context, the current cell state and the neighbouring structure, inserting these codes into the message list. The output module is very simple because it only decodes the next state of the cell, proposed by the activated rule.

### 2.3.3 The evaluation system

When applying a classifier system to a territorial model, the correct definition of an evaluation module is fundamental for building an efficient learning model. In this case the environmental model is a cellular automata, so we run into the problem of evaluating the behaviour of local interaction rules, being able to judge the global emerging behaviour rather than the single action proposed at every temporal step. The evaluation of a configuration generated by the classifier system on the cellular automata, are typically quantitative or qualitative. For a quantitative judgement we can use a measure of spatial entropy, identifying the probability distribution of neighbourhood forms or calculating the fractal dimension of the whole configuration, comparing it with the reference map. A the qualitative judgement can be the opinion of an expert adequately interfaced with the evaluation system, who looks at the evolution of the system and gives a mark, in accordance with what his experience suggests about the possible known developing scenario.

The evaluating system weights the available judgement and to translate them into reward or penalty to distribute among the classifiers contributing to generate the evaluated configuration.

### 2.3.4 The generation of new rules

The genetic system generating new rules is independent of the context to apply a classifier system to a urban model, so we use the genetic process of a standard classifier system.

# 3 A first operative instrument

## 3.1 Running description

The developed system makes it possible, by means of a standard classifier system, given an initial map and a reference map, to draw up the transition rules of the cellular automata. When the expert using the system believes the learning result to be adequate in relation to the modelling requirement, he can suspend the learning procedure and use the obtained rules to simulate the territorial evolution like a normal cellular automata.

The fundamental algorithm of the learning system requires the user to provide an initial map and a reference map and to set the main parameters of the system: the characteristics of the learning system, setting the number of rules to manage and the size of the reward and penalty; the size of taxes: the Head Tax (which subtracts at every cycle a percentage of the strength of each classifier to reduce the strength of never activated rules), the Producer Tax (paid by each classifier able to set out an action, to punish the classifiers proposing many actions but receiving fewer rewards), the Bid Tax (reducing the strength of classifiers proposing overgeneral rules with many don't care #). Having set the learning characteristics the user must set the genetic characteristics, specifying the interval at which apply the genetic algorithms, the number of rules to process, the probability of applying the crossover and mutation algorithm, and the probability of generating don't care type. Finally the interaction radius must be set specifying the number of cells and the sensor structure, available in three kinds:

1. state sensor, the condition part consists of the actual cell state only (if cell state $X \Rightarrow$ next cell state Z)
2. state sensor and majority state in neighbourhood (if cell state $X \wedge$ majority state $Y \Rightarrow$ next cell Z)
3. state sensor, majority in neighbourhood sensor and transportation infrastructures in neighbourhood sensor (if cell state $X \wedge$ majority state $Y \wedge$ there is road $\wedge$ there is railway station $\wedge$ there is a underground station $\Rightarrow$ next cell Z)

where called $S = S_v \cup S_f$ with $S_v$ = variable state set {high and low density housing, industry, commerce, service, vacant} and $S_f$ = fixed state set {parks, railway, public services, railway stations, underground stations, road}, $X \in S_v$, $Y \in S$, $Z \in S_v$.

The learning phase starts, initially choosing at random a rule base, and random choosing a cell (such random choice takes in a noise). If the cell has a fixed state, another is chosen till it gets a variable one. Sensors gives the classifier system back the coded neighbour, the system proposes an action coding the next state; the cell changes state and the system evaluates the adhesion with the reference map, calculated as the number of cells with coincident state; if the adhesion grew, the rule proposing the action gets a reward otherwise it gets a penalty hence increasing or reducing its strength by the amount specified in the learning parameter. If the rule gets a penalty, it is possible to automatically correct the mistake, restoring previous state, so that the system does not have to learn to correct its own mistakes. If the mistake isn't corrected, the system has the additional task to identify rules able to reproduce the reference map and, at the same time, correct the mistake.

When the learning phase is completed, the system evaluates whether the genetic algorithms interval has been used up, so that it starts the genetic algorithms selecting old rules and looking for new rules.

After the application of genetic algorithms, the cycle starts selecting a new cell again until the user stops the procedure.

## 3.2 The software and first simulation

The software is a Windows 3.1 application, built in C++ Borland 4.0 language. we reused the modules developed in the classifiers system built by S. Tanzi and P. Uboldi for the Robotics and Artificial Intelligence Project "ALECSYS" (A Learning Classifiers System) of the Electronic and Information Department of Milan Polytechnic. The software is endowed with an interface to permit the user to act, during the learning process, on each parameter by means of dialogue windows activated through a menu or buttons on a tool bar, buttons representing the main interaction functions like turning on-off the learning and genetic procedures, restarting of the learning procedure with a new rule base, showing the reference map or variation of the elaborating map with respect to the initial map showing the rule base with: birth cycle, condition and action part, strength and kind of generation (random, reproduction, mutation, crossover, partial cover, etc.) of each rule.

The first results prove that the system is able to identify the correct transition rules in a short time with simple learning maps. For example, starting with a map of 100x100 cells of vacant state only and with a reference map of housing state only, the system learns with the first kind of sensors: cell_state(vacant)⇒next_cell_state(housing); if the system has to correct its mistakes: cell_state(any)⇒next_cell_state(housing). In a real example, the learning process is much longer but the identification rules are fairly good. Starting with a map like figure 1, and a reference map like figure 2, the identified rule base, obtained with sensors of the third kind and the application of 888 genetic cycles is shown in table 1. Turning off learning, we obtain the map shown in figure 4, after 1111 cycles. The variation with initial map is shown in figure 5.

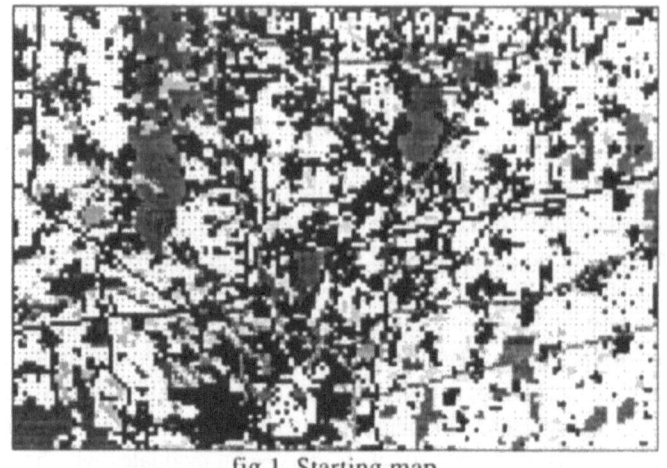

fig 1. Starting map

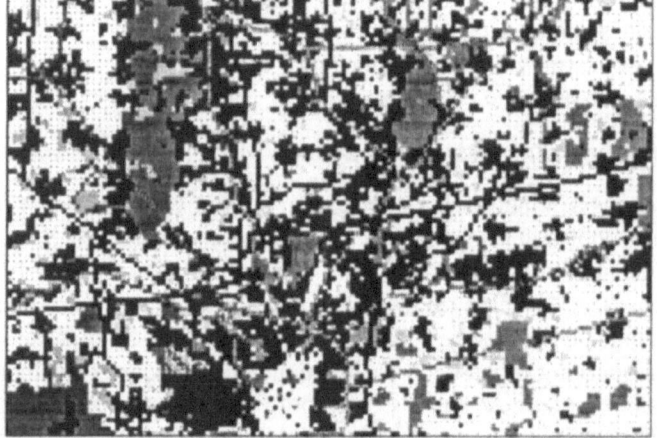

fig. 2 reference map

fig. 3 cells in reference and not in starting map

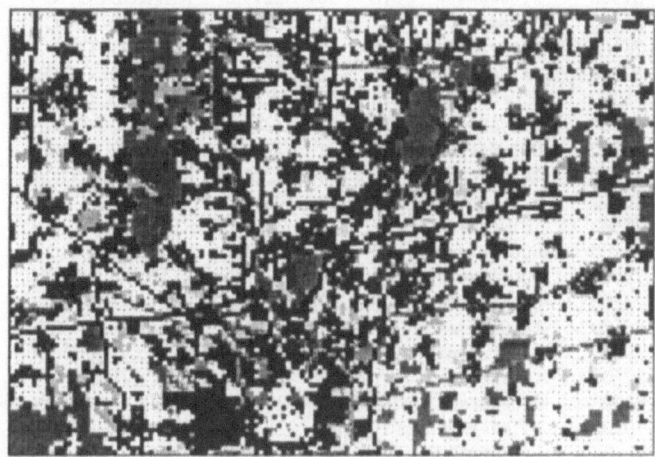

fig. 4 evolution at step 110000

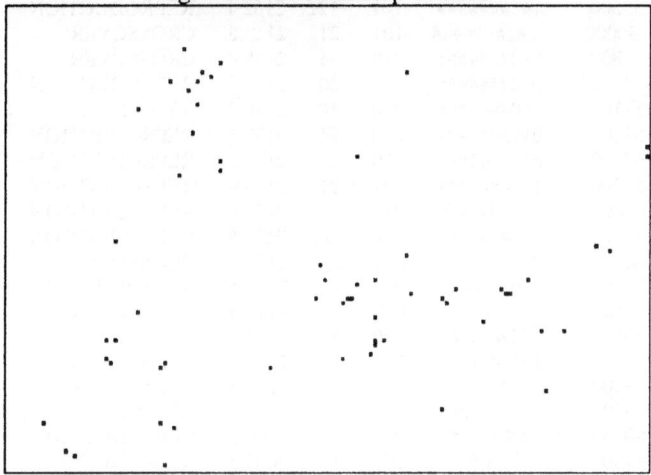

fig. 5 cells in evolution and not in starting map

cycle:        88889

performance:          63.748045

| Number | Id | Born cycle | Condition | Action | Calls | Strength | Born kind |
|---|---|---|---|---|---|---|---|
| 0 | 24476 | 79300 | #10###### | 010 | 44 | 2354.4 | REPRODUCTION |
| 1 | 22678 | 73300 | 010###### | 010 | 71 | 2350.9 | REPRODUCTION |
| 2 | 17892 | 57400 | 000###### | 000 | 118 | 2350.6 | REPRODUCTION |
| 3 | 21977 | 71000 | 001###### | 001 | 70 | 2349.9 | REPRODUCTION |
| 4 | 23743 | 76900 | 000###### | 000 | 60 | 2342.3 | REPRODUCTION |
| 5 | 22487 | 72700 | 000###### | 000 | 68 | 2342.0 | REPRODUCTION |
| 6 | 23406 | 75700 | #10###### | 010 | 49 | 2330.8 | REPRODUCTION |
| 7 | 14447 | 45900 | 010###### | 010 | 176 | 2330.7 | CROSSOVER |
| 8 | 10826 | 33800 | 010###### | 010 | 217 | 2328.7 | REPRODUCTION |
| 9 | 9925 | 30800 | 001###### | 001 | 213 | 2328.7 | REPRODUCTION |
| 10 | 12406 | 39100 | 000###### | 000 | 185 | 2293.9 | REPRODUCTION |
| 11 | 15921 | 50800 | #10###### | 010 | 139 | 2291.5 | REPRODUCTION |

| 12 | 17921 | 57500 | #10####### | 010 | 107 | 2279.6 | REPRODUCTION |
| 13 | 17525 | 56100 | 001####### | 001 | 109 | 2274.1 | REPRODUCTION |
| 14 | 23606 | 76400 | 1010###### | 101 | 46 | 2261.8 | CROSSOVER |
| 15 | 25467 | 82600 | 1#10###### | 101 | 28 | 2237.5 | REPRODUCTION |
| 16 | 24188 | 78300 | 010####### | 010 | 37 | 2225.8 | REPRODUCTION |
| 17 | 21575 | 69600 | #10####### | 010 | 64 | 2224.9 | REPRODUCTION |
| 18 | 24335 | 78800 | 001####### | 001 | 34 | 2224.8 | REPRODUCTION |
| 19 | 25078 | 81300 | 1#1####### | 101 | 28 | 2224.1 | REPRODUCTION |
| 20 | 26584 | 86300 | #10######0 | 010 | 16 | 2219.0 | MUTATION |
| 21 | 23332 | 75500 | 101####### | 101 | 44 | 2216.1 | REPRODUCTION |
| 22 | 11529 | 36100 | 000####### | 000 | 194 | 2212.6 | REPRODUCTION |
| 23 | 24589 | 79700 | 000####### | 000 | 37 | 2209.3 | REPRODUCTION |
| 24 | 7423 | 22500 | 001####### | 001 | 278 | 2195.4 | MUTATION |
| 25 | 26628 | 86500 | 1#10###### | 101 | 13 | 2189.2 | MUTATION |
| 26 | 24434 | 79200 | 1#0##0#0## | 100 | 38 | 2188.5 | REPRODUCTION |
| 27 | 25432 | 82500 | 1#1######0 | 101 | 26 | 2183.5 | MUTATION |
| 28 | 24986 | 81000 | 010####### | 010 | 28 | 2180.6 | REPRODUCTION |
| 29 | 21831 | 70500 | 1#1####### | 101 | 66 | 2172.2 | REPRODUCTION |
| 30 | 24202 | 78400 | 1#1######0 | 101 | 36 | 2166.9 | CROSSOVER |
| 31 | 22462 | 72600 | 010####### | 010 | 64 | 2166.4 | REPRODUCTION |
| 32 | 16748 | 53500 | 1#1####### | 101 | 122 | 2162.4 | REPRODUCTION |
| 33 | 9675 | 30000 | 1#1####### | 101 | 218 | 2159.8 | CROSSOVER |
| 34 | 24026 | 77800 | 010####### | 010 | 35 | 2148.4 | CROSSOVER |
| 35 | 25853 | 83900 | 010####### | 010 | 20 | 2147.8 | REPRODUCTION |
| 36 | 26809 | 87100 | #10#####0# | 010 | 10 | 2140.2 | MUTATION |
| 37 | 25185 | 81700 | 010####### | 010 | 24 | 2137.1 | REPRODUCTION |
| 38 | 19418 | 62400 | #10####### | 010 | 75 | 2133.3 | REPRODUCTION |
| 39 | 19660 | 63300 | 1#1####### | 101 | 91 | 2125.9 | REPRODUCTION |
| 40 | 19777 | 63600 | #10######0 | 010 | 83 | 2123.9 | REPRODUCTION |
| 41 | 10328 | 32100 | #10####### | 010 | 208 | 2121.8 | REPRODUCTION |
| 42 | 18884 | 60700 | 001####### | 001 | 85 | 2121.5 | CROSSOVER |
| 43 | 25454 | 82600 | 000####### | 000 | 22 | 2119.5 | REPRODUCTION |
| 44 | 19991 | 64400 | #10####### | 010 | 89 | 2117.8 | CROSSOVER |
| 45 | 24235 | 78500 | 000#####0 | 000 | 34 | 2113.3 | REPRODUCTION |
| 46 | 25304 | 82100 | 1010##1### | 101 | 23 | 2112.2 | CROSSOVER |
| 47 | 25606 | 83100 | 1#1####### | 101 | 20 | 2110.3 | CROSSOVER |
| 48 | 26567 | 86300 | 1#10###### | 101 | 11 | 2100.9 | MUTATION |
| 49 | 24841 | 80500 | 000####### | 000 | 27 | 2100.3 | REPRODUCTION |
| 50 | 26134 | 84800 | 1#10#0#### | 101 | 15 | 2093.3 | REPRODUCTION |

tab. 1 Example of the 50 best rules selected on 150 after 110000 learning cycle.

## 3.3 Limits and possible developments

Among the limits of the system, the most relevant is in the structure of sensors, due to the need to shorten the code of the condition part to 16 bit, owing to reuse modules of a classifiers system built for a task requiring simpler rules. Among future improvements there is the need to make a through study of the structure of sensors and rule codes, adjusting them to the identification task, request to learn interaction rules in a planning context and improve the evaluation system, adding functions to reward or to punish classifiers after a significant variation of the map and adding more complex measurement systems usingfractal dimension or more explicit urban evaluations.

# Acknowledgments

We wish to thank Prof. M. Colombetti of the Electronic and Information Department of the Milan Polytechnic, for his co-operation and suggestions given us to base the structure of classifier system.

# Bibliography

1. Allen P., Sanglier M. (1981) "Urban evolution, self-organisation and decision making", Environment and Planning, vol. 13
2. Batty M., Xie Y. (1994) "From cells to cities", Environment and Planning B: Planning and Design, vol. 21
3. Batty M., Xie Y. (1995) "An automaton-based explorer for emerging urban forms", The 9th European Colloquium on Theoretical and Quantitative Geography, Spa, Belgium
4. Booker L. B. (1988) "Classifier systems that learn internal world models", Machine Learning n° 3
5. Cecchini A, Viola F. (1992) "Ficties (Fictitious Cities): A simulation for the creation of cities", International Seminar on Cellular Automaton for Regional Analysis, Venezia
6. Cecchini A, Viola F. (1995) "Approaching a generalised urban automata with help! on line (AUGH!)", Workshop on Neural Nets and Urban Cellular Automata, December, Milano.
7. Couclelis H. (1985) "Cellular Worlds: a framework for modelling micro-macro dynamics", Environment and Planning, vol. 17
8. Engelen G., White R., Uljee I., (1995), "Using cellular automata for integrated modelling of socio-environmental systems", Environmental Monitoring and Assessment, vol. 34, Kluver Academic Publishers
9. Ghosh A. (1958) "Input output approach to an allocative system", Economica, 25
10. Grefenstette J. J. (1988) "Credit assignment in rule discovery systems based on genetic algorithms", Machine Learning n° 3
11. Lowry I. S. (1964) "A model of metropolis", Rand Corporation, Santa Monica, California
12. Maniezzo V. (1993) "Algoritmi di apprendimento automatico", Progetto Leonardo, Esculapio Editore, Bologna
13. Phipps M. (1989) "Dynamical behaviour of cellular automata under the constraint of neighbourhood coherence", Geographical Analysis, 21
14. Portugali J., Benenson I. (1995) "Artificial planning experience by means of a euristic", Environment and Planning, vol. 27
15. White R., Engelen G., Uljee I., (1993), "Cellular automata modelling of fractal urban land use patterns: forecasting change for planning applications", Eighth European Colloquium on Theoretical and Quantitative Geography, Budapest.
16. Wilson S.W. (1987) "Classifier systems and the animat problem", Machine Learning n° 2

# A Cooperating Edge Grammar for Edge Detection

G. Adorni

Dipartimento di Ingegneria dell'Informazione, Università di Parma
Viale delle Scienze, 43100, Parma, Italy

V. D'Andrea, R. A. Marques Pereira

Dipartimento di Informatica e Studi Aziendali, Università di Trento
Via Inama 5, 38100, Trento, Italy

## Abstract

Edge detection is a fundamental process in many low-level vision algorithms. It generates a concise and a compact description of the image structure, suitable for manipulation in computer vision tasks. In this paper we describe an application of Cellular Automata to the reconstruction of edges in an image. The discussed approach is an improvement on the original weak membrane model. Such approach showed good performance when implemented on a massively parallel architecture.

## 1 Introduction

An image can be represented as a two dimensional array of pixel values corresponding to the light intensity on the sensor array of a camera. The light intensity in a single point of the scene depends on a large number of factors: color, texture, spatial orientation, distance from the camera and from the light sources, shadows, etc. The light intensity can be encoded by means of several techniques [1]. In this work we will take into account only images in which the light intensity is encoded by means of single values, called grey levels.

The grey level value of a pixel can be modified by many image processing operators following different techniques [2]. A large number of such techniques modify the grey level value of an image pixel according to the values of the neighboring pixels, following a fixed and synchronous rule. The Cellular Automata computational paradigm is naturally suited for describing this kind of computation. This is the main reason for the interest in applying Cellular Automata to image processing.

Image processing problems may require rather heavy computations on a sequential computer. In several applications it is possible to take advantage of parallel and massively parallel machines [3, 4].

The Cellular Automata computational paradigm provides a suitable tool for describing and implementing massively parallel computation, especially for mesh connected or SIMD architectures. These architectures are extremely useful in the field of image processing and, more generally, in computer vision. It is possible to set up a direct mapping between the image pixels, the Cellular Automata cells and the processors of a SIMD array.

In this paper we describe an application of Cellular Automata to edge reconstruction in an image. The discussed approach, is an improvement on the weak membrane method [5, 6], based on a set of rules (the edge grammar) for accepting local edges and a relaxation technique for improving the quality of the result. The complete system showed good performance when implemented on a massively parallel architecture.

# 2 Cooperating processes

Cooperating processes in image processing are parallel local processes that, through iteration, allow information to propagate, obtaining more consistent results in tasks such as image segmentation and edge detection. They are especially useful when it is necessary to assign numerical labels to image parts. The locality of the labeling process can be compensated by *iterating* the labeling process itself, in order to allow the information to propagate. Such an iteration is called a "relaxation" process, due to its resemblance to certain iterative algorithms used in numerical analysis. Very generally, a relaxation process is organized as follows:

- a list of possible labels is independently selected for each image part; and a confidence factor is associated with each possible label;

- the labels (and their confidence factors) of each part are compared with those of related parts, based on a grammatical model of the ideal relationship between labels of adjacent picture parts. Confidence factors are adjusted to reduce inconsistencies. This step can be iterated as many times as required.

During the comparison stage related picture parts are able to communicate and *cooperate*; the different labels that can be assigned to each picture part are in competition, according to the interaction model. As a result, we obtain a labeling consistent with the input data.

A relaxation process is a computational mechanism which allows a set of *myopic* local processes associated with picture parts to interact with each other in order to achieve a globally consistent interpretation of the picture. The goal is to obtain a self assessment of each picture part. This may be represented as a choice from a discrete set of possible labels; plus associating a likelihood factor with each label. The set of labels, or the *universe*, is limited. An example is the range of grey levels that a pixel can assume in an image.

A relaxation process is determined by specifying a model for the neighborhood of a picture part (i.e., the domain of the local relaxation process), a model for the interaction among labels of neighboring picture parts (i.e., the process itself), and a termination condition. The neighborhood model for a relaxation process specifies which pairs of picture parts directly communicate in the process. The neighborhood model is usually designed to establish connections only between *nearby* parts, in order to satisfy the locality constraint. The choice of the set of neighborhood relations will, in general, be determined by the spatial isotropy of the label universe in the image. For example, if we are designing a relaxation process for edge reinforcement, then the relative positions of pixels are crucial since edges are generally aligned, while if we are designing a relaxation process to enhance grey levels, the positional information may be unnecessary.

The interaction model defines how a picture part changes its label based on the labels of its neighbors. An interaction model is composed by:

1. a *representation* of the relationships between labels;

2. a procedure for applying the previously defined representation to change or update labels.

For discrete labels the simplest representation is a list of the set of label pairs that can simultaneously coexist in pairs of neighboring picture parts, while the simplest updating mechanism is a label discarding process. With fuzzy labeling, instead of discarding labels we can modify the likelihood associated to each label by applying an updating procedure.

A wide variety of labeling processes can be used in image processing. Many of these processes operate at pixel level: the parts to be labeled are pixels and the interaction is between pixels. Relaxation processes have proved particularly useful for obtaining the unambiguous labeling of image parts citerosenfeld81a.

The design and control of such processes, however, are not yet well defined. Given a labeling task, how do we choose appropriate neighborhood and interaction models? Given such a process, how many times should it be iterated?

While for discrete processes termination criteria are straightforward to formulate and justify, for a probabilistic relaxation process the situation is more complicated. In practice, such processes can converge to results which are quite poor. Convergence only is not an acceptable criterion, since the limit point may not be an acceptable labeling. Unambiguity (e.g., low entropy) is also not acceptable, since there are too many unambiguous labels, most of which are inconsistent with the given initial labeling.

# 3 The weak membrane model

In contrast with the traditional approaches, which saw edge detection as a thresholding calculus on a previously filtered image, the new generation of image reconstruction schemes combines image smoothing and edge detection in a cooperative and highly effective way.

The *weak membrane model* is a crucial paradigm of modern image reconstruction. The model, constructed on the basis of prior assumptions about universal image structure and expected noise corruption, operates through an iterative process of visual cost optimization, involving edge estimation and edge preserving image smoothing. The construction incorporates some of the standard ideas and techniques from Bayes probability theory, Markov random fields and statistical field theory [5, 6].

The weak membrane model of image reconstruction assumes weakly continuous images corrupted by Gaussian noise. The local non-linear dynamics of the model is based on a competitive interplay between a form of initial data and a characteristic process of anisotropic diffusion. The anisotropy is controlled by the line process $l$ describing the discontinuity contours. The final outcome of a weak membrane process is a restored image and an image of the edges.

From an operational viewpoint, the weak membrane model leads to a massively parallel and highly non linear algorithm of the Cellular Automata type, based on iterative local processing. In the weak membrane case the local up-dating law is not homogeneous, in so far as it depends on the local configuration of an external field which we call the reference field. The processing structure of the weak membrane algorithm is thus an example of Cellular Automaton.

Technically speaking, grey level images are described by image fields, usually denoted by $g$ or $f$. In the weak membrane model, $g$ is used for the image data and $f$ for the image field being transformed by the iterative reconstruction scheme. The main field $f$ can be regarded as a dynamical system $f = f(t)$ with the initial configuration given by the reference field $g = f(0)$. An elastic constrain keeps $f$ close to the image data $g$. The non-linear smoothing acts primarily on the small brightness gradients - neighboring grey level differences below the edge threshold - thereby emphasizing the large brightness gradients.

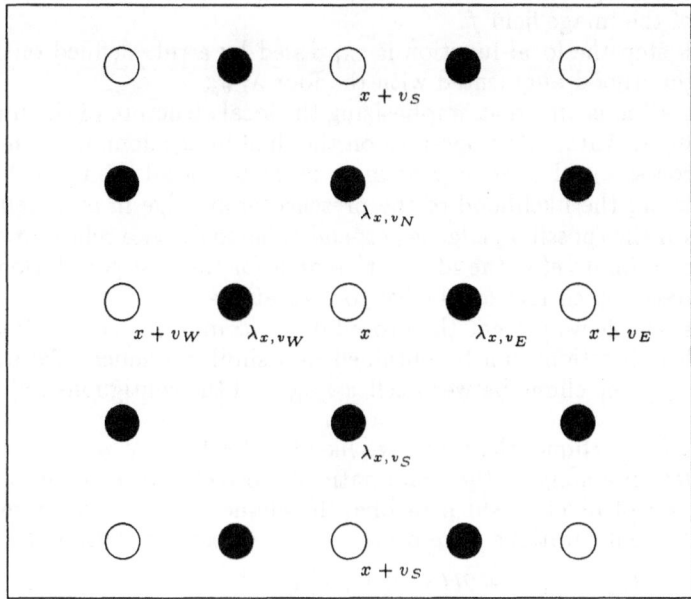

Figure 1: Image domain $L$ (represented with empty circles), and dual image domain $\Lambda$ (full circles).

Consider the image domain $(L)$, represented with empty circles in figure 1, equipped with the usual 1st order neighborhood system, in which pixel $x$ has 4 neighbors $x + v_W, x + v_E, x + v_N, x + v_S$, with adjustments at the image boundaries. Note that the letters N,E,W,S stand for the North, East, West, and South directions. These four directions correspond to the positive and negative translations along the grid directions. For instance, the notation $v_N$ is used to represent a translation in the North direction.

The four binary *cliques* which converge at each pixel $x$ are denoted after their associated positions as: $\lambda_{x,v_N}, \lambda_{x,v_S}, \lambda_{x,v_W}, \lambda_{x,v_E}$. A clique is the connection between two positions in the image domain. For instance, $\lambda_{x,v_N}$ represents the connection between the position $x$ and the position $x + v_N$. This means that: $\lambda_{x,v_N} = \lambda_{x+v_N,v_S}$, and so on.

The set of binary cliques, which we call the *dual image domain* ($\Lambda$), is the effective operational ground of the weak membrane algorithm in the present formulation.

In brief, starting from the image data $f(t = 0) = g$, an iterative process updates the image field $f$ according to some appropriate convergence rate parameter.

The updating law satisfies the Cellular Automata paradigm: the updating function depends only on the neighboring values of $f$ itself and also on those of the reference field $g$ [7].

## 4   Edge grammar

The edge grammar [8] improves the standard weak membrane reconstruction scheme by looking for local linear structures in the line process $l$, before the updating of the image field $f$.

At each step the local function is expressed by a rule defined on Von Neumann neighborhood augmented with the four $\lambda_{x,v_i}$.

The iteration is aimed at emphasizing the local structure of the line process $l$. The Cellular Automaton operates on the dual image domain $\Lambda$ and outputs an edge process $e_\lambda$. The edge process $e_\lambda$ is the set of all the individual values $e(\lambda)$ expressing the likelihood of the presence of an edge in each cell of $\Lambda$. In each cell of $\Lambda$ the (possible) edge is perpendicular to the two cells of $\Lambda$ around it. In order to compute $e(\lambda)$ the idea is to search for the best correlation between the grammatically correct local edge configurations.

As an example we present the procedure for computing $e(\lambda_{x,v_N})$; the values for the other directions can be obtained in a similar manner. We will denote with $\lambda_{x+v_N,v_N}$ the clique between cell $x + v_N$ and the contiguous cell in the $v_N$ direction.

For the $\lambda_{x,v_N}$ clique, the new edge likelihood value $e(\lambda_{x,v_N})$ is given by the best correlation among all the grammatically correct edge configuration in the neighborhood of pixel $x$ which involve the clique $\lambda_{x,v_N}$. Five grammatically correct edge configurations have been defined, whose correlation values are:

$$g_1 = l(\lambda_{x+v_W,v_N}) + 2l(\lambda_{x,v_N}) + l(\lambda_{x+v_E,v_N})$$
$$g_2 = l(\lambda_{x+v_W,v_N}) + l(\lambda_{x,v_N}) + l(\lambda_{x,v_E}) + l(\lambda_{x+v_E,v_S})$$
$$g_3 = l(\lambda_{x+v_W,v_S}) + l(\lambda_{x,v_W}) + l(\lambda_{x,v_N}) + l(\lambda_{x+v_E,v_N})$$
$$g_4 = l(\lambda_{x+v_N,v_W}) + l(\lambda_{x,v_N}) + l(\lambda_{x,v_E}) + l(\lambda_{x+v_E,v_S})$$
$$g_5 = l(\lambda_{x+v_W,v_S}) + l(\lambda_{x,v_W}) + l(\lambda_{x,v_N}) + l(\lambda_{x+v_N,v_E})$$

These five correlation values increase the hypothesis of an edge in $\lambda_{x,V_N}$. They are multiplied by their respective anti-correlations values, related to the presence of an edge in $\lambda_{x,V_N}$. These values are obtained by substituting each line process value with one minus the line process value at the opposite clique:

$$h_1 = 4 - (l(\lambda_{x+v_W,v_S}) + 2l(\lambda_{x,v_S}) + l(\lambda_{x+v_E,v_S}))$$
$$h_2 = 4 - (l(\lambda_{x+v_W,v_S}) + l(\lambda_{x,v_S}) + l(\lambda_{x,v_W}) + l(\lambda_{x+v_E,v_N}))$$
$$h_3 = 4 - (l(\lambda_{x+v_W,v_N}) + l(\lambda_{x,v_E}) + l(\lambda_{x,v_S}) + l(\lambda_{x+v_E,v_S}))$$
$$h_4 = 4 - (l(\lambda_{x+v_N,v_E}) + l(\lambda_{x,v_S}) + l(\lambda_{x,v_W}) + l(\lambda_{x+v_E,v_N}))$$
$$h_5 = 4 - (l(\lambda_{x+v_W,v_N}) + l(\lambda_{x,v_E}) + l(\lambda_{x,v_S}) + l(\lambda_{x+v_N,v_W}))$$

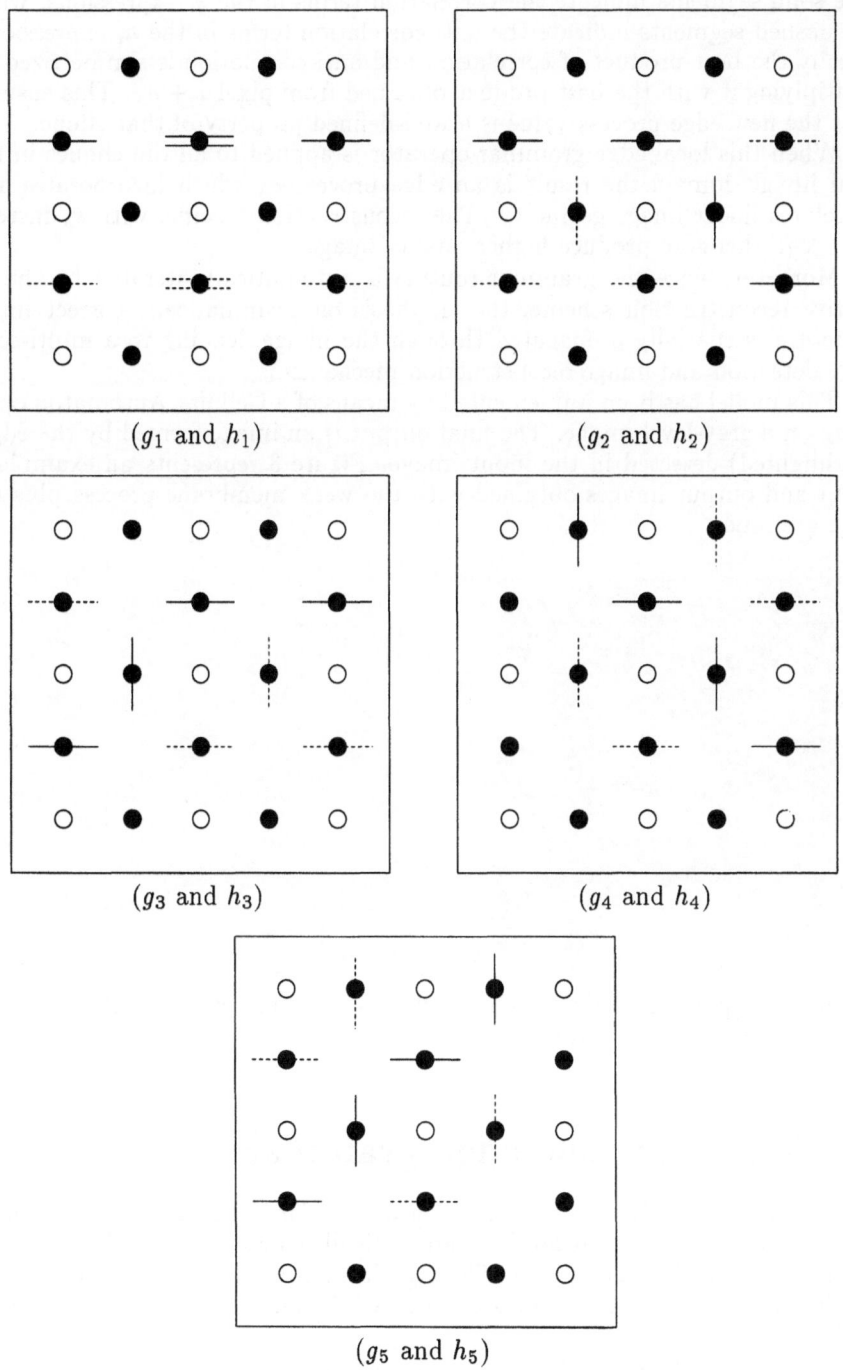

Figure 2: Representation of the rules for $e(\lambda_{x,v_N})$.

The five configurations and their anti-correlations are shown in figure 2. The solid segments indicate the correlation terms in the $g_i$ expressions, while the dashed segments indicate the anti-correlation terms in the $h_i$ expressions. Finally the best product of correlation and anti-correlation is simmetrized by multiplying it with the best product obtained from pixel $x + v_S$. This ensures that the new edge process value is a well-defined property of that clique.

When this local edge grammar operator is applied to all the cliques in the dual image domain the result is an edge process $e_\lambda$ which incorporates and correlates linear image geometry. The reconstruction scheme, with $e_\lambda$ instead of $l_\lambda$, can therefore produce higher quality images.

Moreover, since the grammar routine is automatically iterated by the iterative reconstruction scheme, the emphasis on grammatically correct linear structure is spatially propagated through the image, leading to a multi-scale edge detection and image reconstruction mechanism.

This model has been implemented by means of a Cellular Automaton operating on a grey-level image. The final output is an image formed by the edges (highlighted) detected in the input image. Figure 3 represents an example of input and output images obtained with the weak membrane process plus the edge grammar.

Figure 3: Input image (left) and result image (right) of the weak membrane process plus the edge grammar

# 5 Improving the edge grammar

As we have previously seen, in each clique a new edge likelihood value is given by the best correlation among all the grammatically correct edge configurations in the neighborhood of the pixel $x$. The algorithm performs this task by choosing the highest value of the products of correlation and anti-correlation values for each edge configuration.

The first improvement was to insert a relaxation process before the choice of the highest value. Note that three of the requirements needed to implement a relaxation process are present:

- a set of *labels*: the correlation products,

- a *likelihood* value: the value of the product itself (where the presence of an edge is stronger, the correspondent pixel value will be greater),

- a *neighborhood*: the Von Neumann neighborhood.

There remain to identify:

- a *knowledge representation*: the relations among neighboring edges,

- a *procedure* to change the likelihood values.

The correlation products are computed for each pixel. To create a relaxation process we can compare these values among neighboring pixels. At this level we can use a very simple knowledge representation scheme. Edges tend to *line-up* so:

- if we are considering a *vertical* edge we will compare each pixel with its upper and lower neighbors,

- if we are considering *horizontal* edges we will compare each pixel with its left and right neighbors.

In both cases, if the correlation values calculated for the neighbors are greater than the one present in the *central* pixel, it is reasonable to suppose that the value of the edge in that pixel should be reinforced.

Figure 4: Output of the improved edge grammar.

The edge reinforcement in a certain pixel will result in an increment of its correlation value and, consequently, of its grey level. Since the picture is complemented before being displayed, in the displayed picture we will have darker grey levels where the presence of an edge is stronger. At each step the values of the correlation products are compared with the corresponding values of the neighbors. If the values of both neighbors are greater, then the value of the product is incremented. After updating all the neighborhoods the value of the product for the center cell is replaced by the highest of the neighbors.

A delicate question regards the number of steps needed for each iteration. A first decision can be based on the quality of the input image. Nevertheless, increasing the number of iterations for a noisy image results in a final image that is less consistent with the input data. The observed behavior is that after a small number of steps (depending on the input image) the results converge to a stable configuration. In the example shown in Figure 4, the relaxation process has been iterated five times. Comparing Figure 4 with Figure 3 the improvements due to the relaxation process are self-evident.

# 6   Concluding remarks

The action of the edge grammar, inserted into the Weak Membrane mechanism, was extended by means of a relaxation algorithm, in order to have a consistent labeling in every pixel.

In Figure 5 a small portion of the reference image is expanded to compare the output of the weak membrane process with edge grammar (Figure 3) with that of the weak membrane process with improved edge grammar (Figure 4). We can observe that the improvement is relevant.

 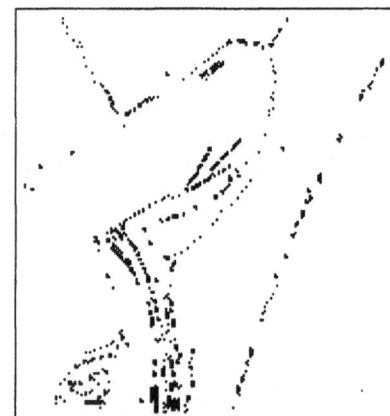

Figure 5: Comparison between the output of the edge grammar (left, magnification of a portion of Figure 3) and the output of the improved edge grammar (right, magnification of the equivalent portion of Figure 4).

From the point of view of the process performances, adding a relaxation process increases the execution time by approximately 10 %.

It must be pointed out that the goal of a relaxation process is the reinforcement of the results obtained; otherwise the process can lead to results that are inconsistent with the problem.

The system has been implemented on an 8K CM-2 Connection Machine. used as an emulator of a Cellular Automata machine. The average execution time for each iteration, including disk I/O, is less then 2 sec.

The authors wish to thank S.Ferrari for his valuable contribution to the experimental tests.

This work has been partially supported by Italian Ministry of Scientific Research under the MURST 40% "Rappresentazione della conoscenza e meccanismi di ragionamento" contract and by the National Research Council (CNR) under the "Progetto Finalizzato Trasporti II" contract.

# References

[1] Dana H. Ballard and Christopher M. Brown. *Computer Vision*. Prentice-Hall, Englewood Cliffs, 1982.

[2] William K. Pratt. *Digital Image Processing*. Wiley, New York, 1978.

[3] Zahid Hussain. *Digital Image Processing: Practical Applications of Parallel Processing Techniques*. Ellis Horwood, New York, 1991.

[4] J. J. Little, G. E. Blelloch, and T. Cass. Algorithmic techniques for computer vision on a fine-grained parallel machine. *IEEE Transaction on Pattern Analysis and Machine Intelligence*, 11(3):244–256, 1989.

[5] Stuart Geman and Donald Geman. Stochastic relaxation, gibbs distribution, and the bayes restoration of images. *IEEE Transaction on Pattern Analysis and Machine Intelligence*, 6:721–741, 1984.

[6] Davi Geiger and Alan L. Yuille. A common framework for image segmentation. *International Journal of Computer VIsion*, 6:227–243, 1991.

[7] Ricardo Alberto Marques Pereira. Early vision with cellular automata fields. In M.S. Garrido and R. Vilela Mendes, editors, *Complexity in Physics and Technology*, pages 193–210. World Scientific, Singapore, 1992.

[8] Ricardo Alberto Marques Pereira and Vincenzo D'Andrea. An edge grammar for edge detection and image reconstruction. Technical Report DII-CE-TR0004-93, Dip. di Ingegneria dell'Informazione, Università di Parma, 1993.

# Hardware Supported Simulation System for Graph Based and 3D Cellular Processing

P. Hartmann, C. Hochberger, R. Hoffmann
R. Schneider, K.-P. Völkmann

Microprogramming and Computer Architecture

Technical University of Darmstadt

Alexanderstr. 10, Darmstadt, D-64283, Germany

### Abstract

This paper presents results of a project in which a hardware supported simulation system for Cellular Processing (CP) is implemented. For three dimensional *regular* CP the hardware architecture, the cellular description language CDL and a method for the efficient generation of the simulator kernel are explained. For *irregular* graph based CP the principal characteristics and their hardware support are discussed.

## 1 Objectives of the project

Cellular Processing (CP) is a generalised term for an inherent massive parallel model of computation where a large number of cells are locally connected to their neighbours. Each cell computes its next state according to a local rule, depending on the neighbours states. The mass of simple parallel state changes results in a complex global state change. Enhancements of the classical and the graph based cellular processing model can be used to create high dimensional and dynamically changing structures in biological, physical or other systems.

The objectives of the project are:

- Development of the theory of classical and graph based CP models.

- Designing a hardware supported simulator, which shall allow real time computation, visualisation and user interaction.

- Designing languages and compilers for an user-friendly description and an efficient simulation.

The work is supported by *Deutsche Forschungsgemeinschaft* (DFG).

## 2 Regular Cellular Processing

### 2.1 Architecture for 3D

In a former project a hardware supported simulation system for 2D regular grids was developed. It uses the implemented special coprocessor *CEPRA-8* [11], containing field programmable gate arrays (FPGA) and digital signal processors (DSP).

A new 3D hardware architecture has been designed, which is scalable in performance and adaptable to different CP models. The implementation will consist of a powerful commercial visualisation board (TMS320C80), a coprocessor board with 16 MB local memory, 3D processing unit and a high speed communication system (PCI). The coprocessor board will be implemented with FPGAs.

The architecture of the 3D processing unit consists of (1) a *xyz-memory* holding one generation, (2) a *xy-plane shifter*, (3) a *line array shifter*, (4) a *computing window* and (5) *p processing elements*.

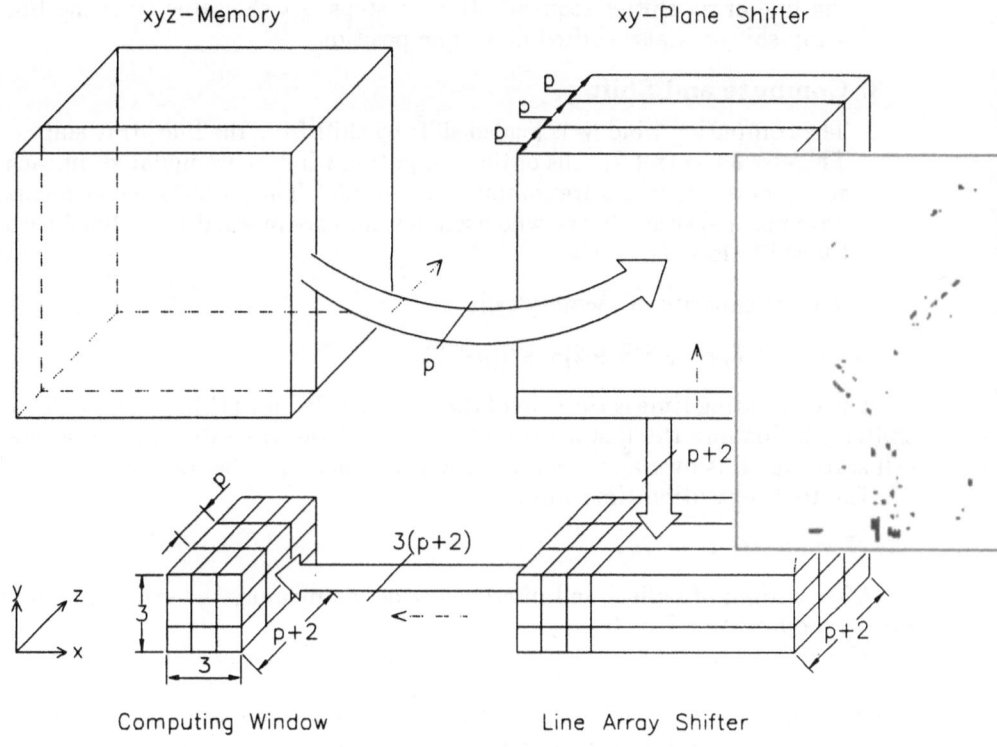

Figure 1: 3D Architecture

The xyz-memory holds $n^3$ cells. $p$ cells in z-axis shall be processed in parallel. The xy-plane shifter consists of 3 z-planes. Each plane holds $n^2p$ cells. The bus between the xyz-memory and the xy-plane shifter is $p$ cell wide.

The line array shifter is of size n in x-axis, 3 in y-axis and $p + 2$ in z-axis. A xz-slice output from the plane shifter can be loaded in $n$ steps into it. It can perform shifts in x direction, thereby loading the computing window. The bus between plane shifter and line array shifter $p + 2$ word wide.

The computing window holds $p$ centre cells (to be computed) in z-axis and all their neighbours. It is of size 3 in x-axis, 3 in y-axis and $p + 2$ in z-axis. It is loaded shift by shift from the line array shifter. The $p$ processing elements have parallel access to the window and are internally pipelined with the shift clock.

The internal operations are:

### 1. Load and Shift Planes

Successive planes are loaded from the memory to the plane shifter, which consist of 3 planes (i=1,2,3). In the start-up cycle all 3 planes are loaded from memory. In the following steps the planes $i$ are shifted to planes $i+1$ and a new memory plane is loaded into the plane 1.

### 2. Load Line Array Shifter

An output slice from the plane shifter of size $n \times 1 \times (p+2)$ is loaded into the line array shifter sequentially in $n$ steps. At the same time the line array shifter is also shifted down one position.

### 3. Compute and Shift

The computing window is loaded shift by shift from the line array shifter. Thereby $3*3*(p+2)$ cells of the computing window are updated. In each step $p$ new cell values are computed in parallel. The $p$ processing elements have parallel access to the whole window and are internally pipelined with the shift clock.

The total capacity of memory cells is

$$C = n^3 + 3pn^2 + 3(p+2)n + 9(p+2).$$

The computing time is the sum of the times needed for (1) loading the plane shifter, (2) loading the line array shifter and (3) the time to compute a new cell state which is overlapped with the writing back into the memory.

The total execution time will result in

$$T = 3n^3/p \cdot t_{CLOCK}.$$

After adding of buffers and interleaving of the memory the execution time can be reduced to $n^3/p \cdot t_{CLOCK}$.

The architecture uses only one central memory. This is possible because the old cell states which are used in the computation are buffered in the xy-plane shifter. It also allows us to add some features to avoid computations of inactive cells [20] and the inhibition of write back of cells which do not change their state.

In the implementation $q$ programmable logical arrays will be used. They emulate $p = q \cdot w/b$ processors, where $w$ is the word length of the buses and $b$ is the scalable number of bits per cell used in the application. Also the number of dimensions and the number of cells per dimensions can be sized to the problem.

*Example.* For an implementation with $q = 8, w = 16, b = 4$ the number of processors will be 32. With $t_{CLOCK} = 32ns$ the number of cell operations per second will be $10^9$.

This architecture is easy scalable in the number of processors and the problem size. Using field programmable logical arrays for the processing elements the required flexibility and performance is reached at relatively low cost and low space.

## 2.2 CDL - A Cellular Description Language

We created the special programming language CDL to specify the local transition (or rule) function for each cell. The usage of a special language has several advantages over classical programming languages:

- Descriptions can be translated to arbitrary target architectures. Among these architectures is the above mentioned family of CEPRA machines. For this purpose the rule is translated into combinational logic. The CDL rule can be used on almost all sequential system (and most of the parallel systems) by transforming it into C code.

- The description of rules in classical programming languages like C or Pascal is very tedious and typically results in deeply nested **if-then-else** statements. CDL contains special functions to simplify testing of complex conditions. These functions operate on groups which are ordered sets of constants e.g. addresses of neighbours. The **one** function can be seen as an $\exists$ quantor, the **all** function as an $\forall$ quantor and the **num** function counts the number of constants in the given group for which the condition holds true. All these functions build implicit loops over the elements of the group.
  On the other hand classical programming languages contain elements that are very hard to translate into hardware. CDL was specifically designed to be translatable to hardware. E.g. all loops are unrolable at compile time.

- CDL allows the user to describe the rule with symbolic state names. It is the compilers task to assign numeric values to the states. This increases the compilers possibilities for optimisations.

- CDL includes a simple, yet powerful visualisation concept. A list of expressions can be defined to specify the colour of a cell under certain conditions. Information on the cell state can be used in the colour expression. This allows a convenient specification of colour ranges.

Existing cellular programming languages like Cellang[3] or Cal did not offer all the advantages mentioned above. Thus we decided to develop a new language. CDL is currently available for two target architectures[16]. Especially the compiler for CEPRA-8L shows that CDL can be used on a wide variety of target architectures.

New extensions of CDL will support moving objects. The common technique for this problem are two phased algorithms. It is very tedious for the user to write such rules. The user has to introduce a lot of new states for each object. The new extensions for CDL shall free the user of this task. Two new sections are introduced into CDL. In the movement section the user specifies where objects want to move and in the conflicts section the resolution of conflicts is specified (e.g. two objects aiming at the same cell). The resulting description can automatically be translated into a CDL description without the extensions, thus producing a two phased model. Whereas on a software simulator the description can be executed in a single phase.

A more detailed description of CDL can be found in [15]. An exhaustive explanation of CDL and several examples can be found at:
http://www.informatik.th-darmstadt.de/~hochberg/CDL

## 2.3   An Abstract High-Level Programming Model

One aim of this project is to describe regular cellular processing in an abstract way. Local transformations, which are responsible for the change of the cell states, will be described in CDL. All characteristics, which influence the processing of cellular models on the global transformation level, will be included into an abstract formalism. This formalism is the starting point from which the target code of an individual cellular processing kernel (CP kernel) for the target architecture will be derived.

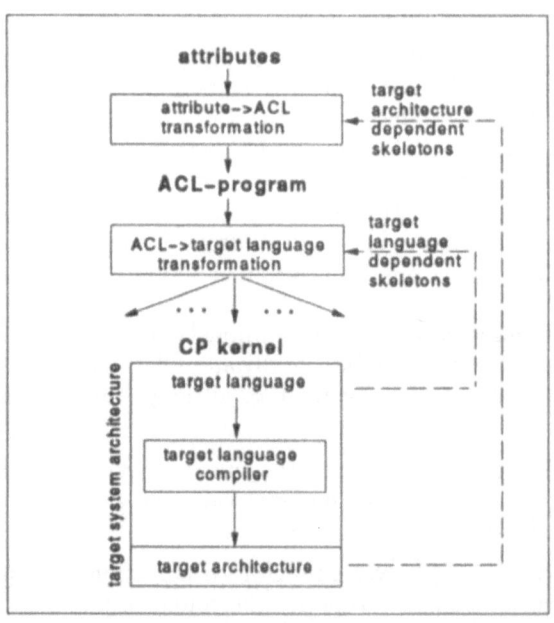

Figure 2: The ACL traversion steps and CP kernel generation

CP kernels are the main part of a simulation system and organise the systematic update of the cell states for every generation. The CP kernel is responsible for the global transformation of the cellular model. With ACL optimised CP kernels can be generated taking into account the attributes of the cellular processing model:

The **algorithmic behaviour** of the model, which includes the *synchronous* and *asynchronous* global transformation of the cellular field. In case of the asynchronous transformation *probabilistic* and *n-phased* global transformations are supported. Examples for a 2-phased global transformation are the "chess board pattern" and the Margolus-model with a special neighbourhood type. Algorithmic behaviour also includes the *edge handling* of the model, like a linear or cyclic wrapped cellular field.

The **data structure** of a cellular processing model may be complex compared to the standard models in [21]. Therefore multidimensional grids and special neighbourhoods for crystal lattices are supported.

**Optimisation strategies** in both space and time are needed to map cellular processing models onto dedicated target architectures. Optimisation is included on the target code level of the CP kernel and in the global transformation algorithm. To further optimise the simulation process additional mechanism are provided for the visualisation of complex multi state 3D-models.

The **underlying target architecture** is included in the process of the construction of an individual CP kernel on target code level. Depending on the features of the dedicated hardware special CP kernel codes are generated. This process is based on abstract program skeletons, which are substituted to form the final target code.

This amount of highly-focused requirements lead to a formalism, which will be expressed using the newly introduced ACL technique [19] (Fig. 2). ACL (*Abstract Cellular model Language*) is not only an abstract language, which describes the cellular attributes, but it is also a high-level programming environment for generating ready-to-run CP kernels on the fly. Therefore ACL is based on cellular attributes and substitutes abstract program skeletons, which may be specified for classes of target architectures. The resulting CP kernels are compilable and linkable with standard compiling techniques.

ACL is based on the idea of a mean to construct CP kernels with embedded application specific optimisation strategies, which guarantee an efficient simulation of cellular models [1]. Since one single universal CP kernel with various mechanism can not fulfil the application specific tasks, a new way is gained through a *rapid prototyping* approach. In combination with special cellular programming languages this tool will simplify the creation of CP kernels for parallel systems and optimise CP kernels for sequential processors.

# 3 Graph Based Cellular Processing

Especially biological cellular tissues are complex systems which give rise to a multitude of sophisticated questions, for instance (1) *how do cells cooperate to perform global tasks*, (2) *how do cells coordinate cell proliferation, cell death and cell migration*, or even (3) *how can a set of non-differentiated cells give rise to the development of a complex organism* ? Analytical methods are limited to systems of little complexity, and experimental investigations cannot help to explain complex interactions between a multitude of cells. If an organism is interpreted as a complex cellular system, methods from the field of cellular automata can be applied. In order to avoid artifacts coming from the abstraction when constructing the model (i.e. because of the underlying static and regular grid), the CA paradigm needs to be adapted to this class of problems.

A necessary generalisation of cellular automata means to introduce *structural dynamics* of the underlying mesh. This implicitly means that the mesh is *inhomogeneous*, too. On the theoretical level we use *cellular hypergraphs* for modelling the spatial structure and *replacement systems* for describing the temporal dynamics of the cells [9]. A replacement system consists of a *set of productions* where each production describes a valid modification of the graph's structure. An important feature of such a replacement system is that its productions can operate *in parallel*, and consequently one of the major characteristics of cellular automata, their inherently massive-parallel operation mode, is preserved.

Our approach has no limitations according to the *dimensions* of the model as well as the *type of productions*, which is an advantage in comparison to other existing theories (see [17, 6]). And it allows us not only to describe and simulate the temporal development of three-dimensional structures like plant tissues, but even of the more complicated tissues of animals. A variety of additional applications can be found in *adaptive mesh generation* for finite element methods, *computer graphics, neural networks* and *spatial databases*.

Although graph-like structures constitute a convenient method for dealing with a multitude of problems, new questions arise which need to be solved, i.e.: (1) *how can we navigate through (especially three-dimensional) graph-based cellular structures*, (2) *does there exist an efficient subgraph matching algorithm (for testing the applicability of graph productions)*, and (3) *can we support the addressing of graph components and the application of graph productions with hardware* ?

Questions (1) and (2) could be solved introducing a concept for addressing components and for navigating through cellular hypergraphs [10]. This addressing model can be used to formulate arbitrary complex addressing expressions without the need to refer to a global coordinate system. Hence it constitutes the basis for the formulation of the local transition function. And the subgraph search, which is the most time consuming step when running a replacement system, can be performed in *linear time* if a pre-compiled search strategy consisting of a sequence of addressing steps is used [9].

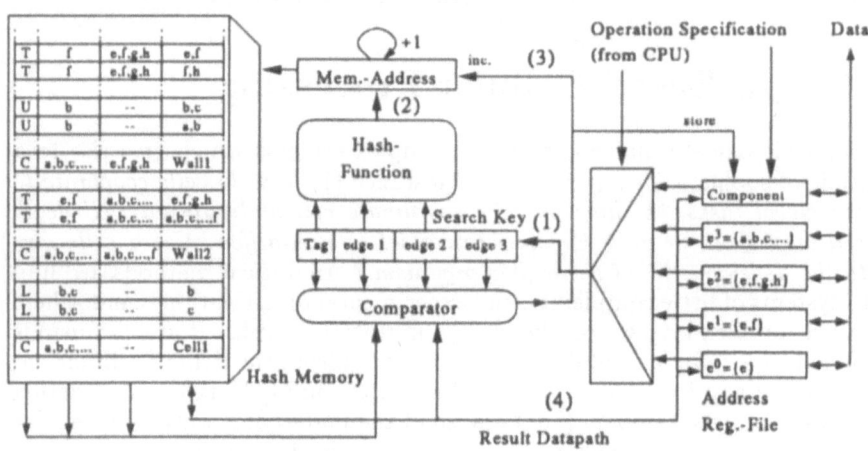

Figure 3: Architecture of a coprocessor for graph-based cellular structures.

Question (3) required the development of a data structure which can be implemented efficiently. For storing cellular hypergraphs a data structure is used where a set of *tuples* expresses relationships between different components of the graph in a unified way (instead of a conventional list based data structure). This data structure can be supported by a coprocessor which stores and administers the graph while the host triggers graph operations and performs the computations of associated state values (see figure 3).

The *register file* is used to point to edges of the graph while navigating through the cellular structure. Different tuples in the *hash memory* represent the relationships between edges (tag *U* for *upper* edge, *L* for *lower* edge), support relative addressing steps (*T* for *address triple*) and store associations between the graph and its cellular components (tag *C*). All queries to the graph can be performed using a unified scheme where (1) a search key is compiled according to the operation specification, (2) the hash function is computed and the starting address for the search is loaded into the sequence register, (3) the sequence register is incremented until the desired tuple is found, and (4) the result is stored in the related register.

# References

[1] M. Cannataro, S. Di. Gregorio, R. Rongo, W. Spataro, G. Spezzano and D. Talia. *A Parallel Cellular Automata Environment on Multicomputers for Computational Science.* Parallel Computing 21 (1995)

[2] K. M. Decker, J. J. Dvorak and R. M. Rehmann. *A Knowledge-Based Scientific Parallel Programming Environment.* Swiss Scientific Computing Center, CSCS-TR-93-07, Manno 1993

[3] J.D. Eckart. *A Cellular Automata Simulation System.* in: SIGPLAN Notices **26**:80–85, August 1991

[4] H. Gutowitz. *Cellular Automata - Theory and Experiment.* MIT Press, Cambridge, Massachusetts, 1991

[5] M. M. Gutzmann and St. Kindermann. *Transformation based Development of Efficient Programs for Massively Parallel Architectures*, Dep. of Computer Science, University of Jena/Erlangen, Germany

[6] P. Prusinkiewicz and A. Lindenmayer. *The Algorithmic Beauty of Plants*, Springer-Verlag 1990

[7] P. Hartmann. *Efficient Subgraph Matching within Cellular Hypergraphs*, Second International Conference on Developments in Language Theory, Magdeburg, Germany, 1995

[8] P. Hartmann. *Parallel and Distributed Processing of Cellular Hypergraphs*, Third International Conference on Parallel Computing Technologies (PaCT-95), St. Petersburg, Russia, 1995

[9] P. Hartmann. *Implementation of Parallel Replacement Systems for Cellular Hypergraphs*, in: Journal of Automata, Languages and Combinatorics 1996 vol. 1, Nr. 2, pp. 129–146, Otto-von-Guericke-Universität Magdeburg

[10] P. Hartmann. *Adressing of Components Within Inhomogeneous Cellular Structures*, in: Proceedings of the VIIth International Workshop on Parallel Processing by Cellular Automata and Arrays (PARCELLA'96) September 96, Mathematical Research, pp. 38–46, Akademie Verlag

[11] R. Hoffmann and K.-P. Völkmann. *CEPRA-8: A Cellular Processing Machine.* in: The Proc. of the 2nd Intern. Workshop for Massive Parallelism, Capri, Italy, October 1994

[12] R. Hoffmann, K.-P. Völkmann, M. Sobolewski. *The Cellular Processing Machine CEPRA-8. in:* Proceedings of the VIth International Workshop on Parallel Processing by Cellular Automata and Arrays, Potsdam, Germany, September 1994

[13] R. Hoffmann. *Cellular Processing Architectures*, Mitteilungen - Gesellschaft für Informatik e.V., Parallel-Algorithmen und Rechnerstrukturen, PARS Workshop Dresden, Germany, April 1993

[14] R. Hoffmann. *Rechnerentwurf: Rechenwerke, Mikroprogramierung, RISC*, Oldenbourg Verlag München, 1993

[15] C. Hochberger and R. Hoffmann. *CDL - A Language for Cellular Processing in:* Proc. Second International Conference on Massively Parallel Computing Systems (MPCS), Ischia, Italy, 1996

[16] C. Hochberger, R. Hoffmann and S. Waldschmidt. *Compilation of CDL for Different Target Architectures. in:* Parallel Computing Technologies (LNCS 964), Springer Verlag, 1995

[17] A. Lindenmayer. *Models for plant tissue development with cell division orientation regulated by preprophase bands of microtubules*, volume 10 of *Differentiation*, pages 1–10, Springer-Verlag, 1984

[18] N. Margolus. *CAM-8: a computer architecture based on cellular automata.*, MIT Lab. for Computer Science, Cambridge Mass. 01239, December 1993

[19] R. Schneider. *ACL - A High-Level Programming Model for Cellular Processing with Optimization Strategies. in:* Proc. Workshop on High-Level Programming Models and Supportive Environments (HIPS), Honolulu, Hawaii, April 1996

[20] R. Schneider, R. Hoffmann. *An Optimization Strategy for Cellular Processing.*, *in:* Proceedings MPCS'96, Ischia, Italy, May 1996

[21] T. Toffoli, N. Margolus. *Cellular Automata Machines.* MIT Press, Cambridge, Massachusetts, 1987

[22] S. Wolfram. *Cellular Automata and Complexity.* Addison-Wesley, Reading, Massachusetts, 1994

# Evolvable Cellular Machines

## Moshe Sipper

Logic Systems Laboratory, Swiss Federal Institute of Technology,
1015 Lausanne, Switzerland. E-mail: Moshe.Sipper@di.epfl.ch

## Marco Tomassini

Institute of Computer Science, University of Lausanne, and
Logic Systems Laboratory, Swiss Federal Institute of Technology,
1015 Lausanne, Switzerland. E-mail: Marco.Tomassini@di.epfl.ch

### Abstract

A major impediment preventing ubiquitous computing with cellular automata (CA) stems from the difficulty of utilizing their complex behavior to perform useful computations. In this paper *non-uniform* CAs are studied, presenting the *cellular programming* algorithm for co-evolving such CAs to perform computations. The algorithm's efficacy is demonstrated on two non-trivial computational tasks, namely synchronization and random number generation; furthermore, we present initial results demonstrating the robustness of our evolved systems. We believe that cellular programming holds potential for attaining 'evolving ware', *evolware*, which can be implemented in software, hardware, or other possible forms, such as bioware.

## 1   Introduction

Cellular automata (CA) are dynamical systems in which space and time are discrete, exhibiting three notable features: massive parallelism, locality of cellular interactions, and simplicity of basic components (cells). A major impediment preventing ubiquitous computing with CAs stems from the difficulty of utilizing their complex behavior to perform useful computations. Designing CAs to have a specific behavior or perform a particular task is highly complicated, thus severely limiting their applications; automating the design (programming) process would greatly enhance the viability of CAs [Mitchell *et al.*, 1994]. A prime motivation for studying CAs stems from the observation that they are naturally suited for hardware implementation, with the potential of exhibiting extremely fast and reliable computation that is robust to noisy input data and component failure [Gacs, 1985].

Recent studies have shown that CAs can be evolved, using genetic-algorithm based methods, to perform non-trivial computational tasks. The model investigated in this paper is an extension of the CA model, termed *non-uniform cellular automata* [Sipper, 1994, Vichniac *et al.*, 1986]. Such automata function in the same way as uniform ones, the only difference being in the cellular rules that need not be identical for all cells. Our main focus is on the *evolution* of non-uniform CAs to perform computational tasks, employing a local, co-evolutionary algorithm, an approach referred to as *cellular programming*. In this paper we present our approach and demonstrate its application to two non-trivial computational problems: synchronization and random number generation. We believe that cellular programming holds potential for attaining 'evolving ware', *evolware*, which can be implemented in software, hardware, or other possible forms. Of particular interest is the issue of evolving hardware, which has recently made its appearance on the artificial evolution scene [Sanchez and Tomassini, 1996].

The application of genetic algorithms to the *evolution* of *uniform* cellular automata was initially studied by [Packard, 1988] and recently undertaken by the EVCA (evolving CA) group [Mitchell *et al.*, 1994, Das *et al.*, 1995], demonstrating that CAs can be evolved to perform computational tasks. They carried out experiments involving uniform, one-dimensional CAs with $k = 2$ and $r = 3$, where $k$ denotes the number of possible states per cell and $r$ denotes the radius of a cell, i.e., the number of neighbors on either side (thus each cell has $2r + 1$ neighbors, including itself). Spatially periodic boundary conditions are used, resulting in a circular grid. We had first studied non-uniform CAs in [Sipper, 1994, Sipper, 1995b] and demonstrated in [Sipper, 1995a] that universal computation can be attained in such CAs. The universal systems we presented are simpler than previous ones and are *quasi*-uniform, meaning that the number of distinct rules is extremely small with respect to rule space size; furthermore, the rules are distributed such that a subset of dominant rules occupies most of the grid. The co-evolution of non-uniform, one-dimensional CAs to perform computations was undertaken in [Sipper, 1996a], where the cellular programming algorithm was presented; we showed that high performance, non-uniform CAs can be co-evolved not only with radius $r = 3$, as previously studied, but also for smaller radiuses, most notably for minimal $r = 1$. It was also found that evolved systems exhibiting high performance are quasi-uniform.

The cellular programming algorithm is delineated in the next section. In Section 3, we demonstrate its application to two computational tasks, namely synchronization and random number generation. Our conclusions are presented in Section 4.

## 2    The cellular programming algorithm

We study 2-state, non-uniform CAs, in which each cell may contain a different rule. A cell's rule table is encoded as a bit string, known as the "genome", containing the next-state (output) bits for all possible neighborhood configurations,[1] listed in lexicographic order; e.g., for CAs with $r = 1$, the genome consists of 8 bits, where the bit at position 0 is the state to which neighborhood configuration 000 is mapped to and so on until bit 7 corresponding to neighborhood configuration 111. Rather than employ a *population* of evolving, uniform CAs, as with genetic algorithm approaches, our algorithm involves a *single*, non-uniform CA of size $N$. Cell rules are initialized at random, uniformly distributed among

---

[1]The term 'configuration' refers to an assignment of states to grid cells.

different fractions of output 1 bits. Initial configurations are then generated at random, in accordance with the task at hand. For each initial configuration the CA is run for $M$ time steps. Each cell's *fitness* is accumulated over $C = 300$ initial configurations, where a single run's score is 1 if the cell is in the correct state after $M$ iterations, and 0 otherwise. After every $C$ configurations evolution of rules occurs by applying crossover and mutation. This evolutionary process is performed in a completely *local* manner, where genetic operators are applied only between directly connected cells. It is driven by $nf_i(c)$, the number of fitter neighbors of cell $i$ after $c$ configurations. The pseudo-code of our algorithm is delineated in Figure 1.

```
for each cell i in CA do in parallel
    initialize rule table of cell i
    fᵢ = 0 { fitness value }
end parallel for
c = 0 { initial configurations counter }
while not done do
    generate a random initial configuration
    run CA on initial configuration for M time steps
    for each cell i do in parallel
        if cell i is in the correct final state then
            fᵢ = fᵢ + 1
        end if
    end parallel for
    c = c + 1
    if c mod C = 0 then { evolve every C configurations}
        for each cell i do in parallel
            compute nfᵢ(c) { number of fitter neighbors }
            if nfᵢ(c) = 0 then rule i is left unchanged
            else if nfᵢ(c) = 1 then replace rule i with the fitter neighboring rule,
                    followed by mutation
            else if nfᵢ(c) = 2 then replace rule i with the crossover of the two fitter
                    neighboring rules, followed by mutation
            else if nfᵢ(c) > 2 then replace rule i with the crossover of two randomly
                    chosen fitter neighboring rules, followed by mutation
                    (this case can occur if the cellular neighborhood includes
                    more than two cells)
            end if
            fᵢ = 0
        end parallel for
    end if
end while
```

Figure 1: Pseudo-code of the cellular programming algorithm.

The genetic operators of crossover and mutation are those used in genetic algorithms [Mitchell, 1996]. Crossover between two rules is performed by selecting at random (with uniform probability) a single crossover point and creating a new rule by combining the first rule's bit string before the crossover point with the second rule's bit string from this point onward. Mutation is applied to the bit string of a rule with probability 0.001 per bit.

There are two main differences between our algorithm and the standard genetic algorithm: (a) A standard genetic algorithm involves a population of evolving, uniform CAs; all CAs are ranked according to fitness, with crossover occurring between *any* two individuals in the population. Thus, while the CA runs in accordance with a local rule, evolution proceeds in a *global* manner. In contrast, our algorithm proceeds *locally* in the sense that each cell has access only to its locale, not only during the run but also during the evolutionary phase, and no global fitness ranking is performed. (b) The standard genetic algorithm involves a population of *independent* problem solutions; each CA is run independently, after which genetic operators are applied to produce a new population. In contrast, our CA *co-evolves* since each cell's fitness depends upon its evolving neighbors.

This latter point comprises a prime difference between our algorithm and parallel genetic algorithms, which have attracted attention over the past few years. These aim to exploit the inherent parallelism of evolutionary algorithms, thereby decreasing computation time and enhancing performance [Tomassini, 1995]. A number of models have been suggested, among them coarse-grained, island models [Starkweather *et al.*, 1991, Cohoon *et al.*, 1987, Tanese, 1987], and fine-grained, grid models [Tomassini, 1993, Manderick and Spiessens, 1989]. The latter resemble our system in that they are massively parallel and local; however, the co-evolutionary aspect is missing. As we wish to attain a system displaying global computation, the individual cells do not evolve independently as with genetic algorithms (be they parallel or serial), i.e., in a "loosely-coupled" manner, but rather co-evolve, thereby comprising a "tightly-coupled" system.

# 3 Results

In this section we demonstrate the application of our algorithm to two non-trivial computational problems, namely synchronization and random number generation. The cellular space used is minimal, with $k = 2$ and $r = 1$. Performance values reported hereafter represent the average fitness of all grid cells after $C$ configurations, normalized to the range $[0, 1]$.

## 3.1 The synchronization task

The one-dimensional synchronization task was introduced by [Das *et al.*, 1995] and studied by us in [Sipper, 1996b] using non-uniform CAs. In this task the CA, given any initial configuration, must reach a final configuration, within $M$ time steps, that oscillates between all 0s and all 1s on successive time steps. It belongs to a class of problems studied in other domains, such as distributed computing, known as firing squad problems [Lamport and Lynch, 1990].

The task is non-trivial since synchronous oscillation is a global property of a configuration, whereas a small radius CA employs only local interactions. Thus, while local regions of synchrony can be directly attained, it is more difficult to design CAs in which spatially distant regions are in phase. Since out-of-phase regions can be distributed throughout the lattice, propagation of information must occur over large space-time distances (i.e., $O(N)$) to remove these phase defects and produce a globally synchronous configuration [Das *et al.*, 1995].

In [Sipper, 1996b] we studied non-uniform, one-dimensional, minimal radius $r = 1$ CAs of size $N = 149$. The size of *uniform*, $r = 1$ CA rule space is small, consisting of only $2^8 = 256$ rules. This enabled us to check each and every one of these rules on the synchronization task, a feat not possible for larger values of $r$. Our results show that the maximal performance for uniform, $r = 1$ CAs is 0.84. For the cellular programming algorithm we used randomly generated initial configurations, with the CA being run for $M = 150$ time steps. We found that quasi-uniform CAs had co-evolved that exhibit near-perfect performance, thereby surpassing any possible uniform CA. Figure 2a depicts the operation of a co-evolved CA, along with a rules map, depicting the distribution of rules by assigning a unique color to each distinct rule. A detailed investigation of the one-dimensional synchronization task can be found in [Sipper, 1996b].

We have recently begun an investigation of the robustness of the solutions discovered by evolution. Toward this end we consider the effects of faults on the CA's behavior with the intention of studying the recovery capabilities of the system. We focus on one type of error where a cell updates its state in a non-deterministic manner: at each time step, the cell's next state is that specified in the rule table, with probability $1 - p_f$, or the complementary one with probability $p_f$; $p_f$ is denoted the *fault probability*, representing the probability that a cell will incorrectly update its state. Figures 2b and 2c demonstrate the effects of different $p_f$ values on the CA's behavior. We note that for small $p_f$ values quick recovery is possible, while for larger values of $p_f$ behavior becomes more erratic. We are currently conducting a quantitative study of the fault-tolerance issue.

## 3.2   Random number generation

Random numbers are needed in a variety of applications, yet finding good random number generators, or randomizers, is a difficult task [Park and Miller, 1988]. To generate a random sequence on a digital computer, one starts with a fixed length seed, then iteratively applies some transformation to it, progressively extracting as long as possible a random sequence. Such numbers are usually referred to as *pseudo*-random, as distinguished from true random numbers resulting from some natural physical process. In order to demonstrate the efficiency of a proposed generator, it is usually subjected to a battery of empirical and theoretical tests, among which the most well known are those described in [Knuth, 1981].

In the last decade CAs have been used to generate random numbers. The first such work is that of [Wolfram, 1986], in which rule 30 is extensively studied for its ability to produce random, temporal bit sequences.[2] Such sequences are obtained by sampling the values that a particular cell attains as a function of time. In [Wolfram, 1986] the uniform, two-state, $r = 1$, rule 30 CA is initialized with a configuration consisting of a single cell in state 1, with all other cells being in state 0; the initially non-zero cell is the site at which

---

[2]Rule numbers are given in accordance with Wolfram's convention [Wolfram, 1983], representing the decimal equivalent of the binary number encoding the rule table.

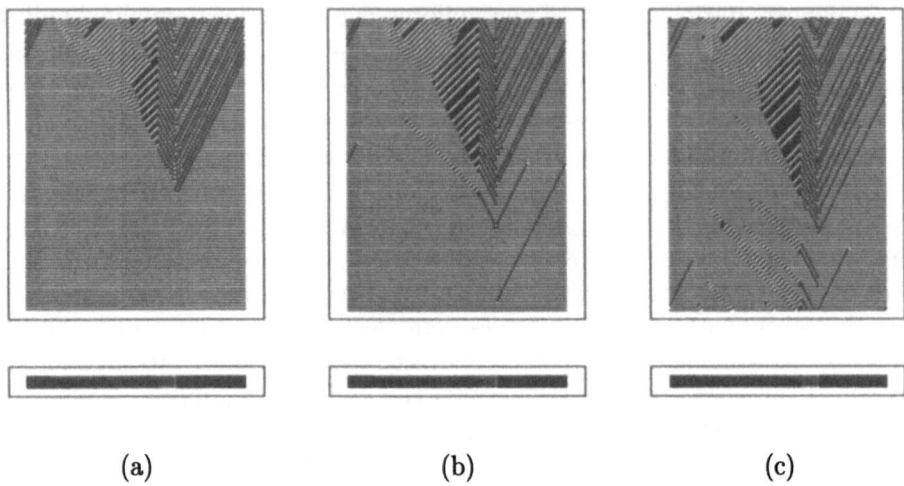

Figure 2: The one-dimensional synchronization task: Operation of a co-evolved, non-uniform, $r = 1$ CA. Grid size is $N = 149$. White squares represent cells in state 0, black squares represent cells in state 1. The pattern of configurations is shown through time (which increases down the page). Initial configurations were generated at random. Top figures depict space-time diagrams, bottom figures depict rule maps. (a) $p_f = 0$. (b) $p_f = 0.0001$. (c) $p_f = 0.001$.

the random temporal sequence is generated. Wolfram studied this particular rule extensively, demonstrating its suitability as a high-performance randomizer which can be efficiently implemented in parallel; indeed, this CA is one of the standard generators of the massively parallel Connection Machine CM2 [Thi, 1991]. A non-uniform CA randomizer was presented by [Hortensius et al., 1989a, Hortensius et al., 1989b] (based on the work of [Pries et al., 1986]), consisting of two rules, 90 and 150, arranged in a specific order in the grid. The performance of this CA in terms of random number generation was found to be at least as good as that of rule 30, with the added benefit of less costly hardware implementation. It is interesting in that it combines two rules, both of which are simple linear rules that do not comprise good randomizers, to form an efficient, high-performance generator. An example application of such CA randomizers has recently been demonstrated by [Chowdhury et al., 1995] who designed a low-cost, high-capacity associative memory.

An evolutionary approach for obtaining random number generators was taken by [Koza, 1992], who applied genetic programming to the evolution of a symbolic LISP expression that acts as a rule for a uniform CA (i.e., the expression is inserted into each CA cell, thereby comprising the function according to which the cell's next state is computed). He demonstrated evolved expressions that are equivalent to Wolfram's rule 30. The fitness measure used by Koza is the entropy $E_h$: let $k$ be the number of possible values per sequence position (in our case CA states) and $h$ a subsequence length. $E_h$ (measured in bits) for the set of $k^h$ probabilities of the $k^h$ possible subsequences of length $h$ is given by:

$$E_h = -\sum_{j=1}^{k^h} p_{h_j} \log_2 p_{h_j}$$

where $h_1, h_2, \ldots, h_{k^h}$ are all the possible subsequences of length $h$ (by convention, $\log_2 0 = 0$ when computing entropy). The entropy attains its maximal value when the probabilities of all $k^h$ possible subsequences of length $h$ are equal to $1/k^h$; in our case $k = 2$ and the maximal entropy is $E_h = h$. Koza evolved LISP expressions which act as rules for uniform, one-dimensional CAs. The CAs were run for 4096 time steps and the entropy of the resulting temporal sequence of a designated cell (usually the central one) was taken as the fitness of the particular rule (i.e., LISP expression). In his experiments Koza used a subsequence length of $h = 4$, obtaining rules with an entropy of 3.996. The best rule of each run was re-tested over 65536 time steps, some of which exhibited the maximal entropy value of 4.0.

For the cellular programming algorithm the cell's fitness score for a single configuration is defined as the entropy $E_h$ of the temporal sequence, after the CA has been run for $M$ time steps; $f_i$ is then updated as follows (refer to Figure 1):

**for** each cell $i$ **do in parallel**
$\qquad f_i = f_i +$ entropy $E_h$ of the temporal sequence of cell $i$
**end parallel for**

Rather than restrict ourselves to one designated cell, we consider all grid cells, thus obtaining $N$ random sequences in parallel, rather than a single one. Initial configurations for our evolving, non-uniform CA are selected at random,[3] after which the CA is run for $M = 4096$ time steps. In our simulations (using grids of sizes $N = 50$ and $N = 150$), we observed that the average cellular entropy taken over all grid cells is initially low (usually in the range $[0.2, 0.5]$), ultimately evolving to a maximum of 3.997, when using a subsequence size of $h = 4$ (i.e., entropy is computed by considering the occurrence probabilities of 16 possible subsequences, using a "sliding window" of length 4).

We performed several such experiments using $h = 4$ and $h = 7$; the evolved, non-uniform CAs attained average fitness values (entropy) of 3.997 and 6.978, respectively. We then re-tested our best CAs over $M = 65536$ times steps (as in [Koza, 1992]), obtaining entropy values of 3.9998 and 6.999, respectively. Interestingly, when we performed this test with $h = 7$ for CAs which were evolved using $h = 4$, high entropy was displayed as for CAs which were originally evolved with $h = 7$. The entropy results are comparable to those of [Koza, 1992] as well as to the rule 30 CA of [Wolfram, 1986] and the non-uniform, rules $\{90, 150\}$ CA of [Hortensius et al., 1989a, Hortensius et al., 1989b]. Note that while our fitness measure is local, the evolved entropy results reported above represent the average of all grid cells; thus, we obtain $N$ random sequences in

---

[3]A standard, 48-bit, linear congruential algorithm proved sufficient for the generation of initial configurations.

parallel, rather than a single one. Figure 3 demonstrates the operation of three CAs discussed above: rule 30, rules {90, 150}, and a co-evolved CA. A more detailed investigation has been carried out in [Sipper and Tomassini, 1996b, Sipper and Tomassini, 1996a], using tests described in [Knuth, 1981], suggesting that good randomizers can be evolved; these exhibit behavior at least as good as that of previously described CA generators, with notable advantages arising from the existence of a "tunable" algorithm for the generation of randomizers.

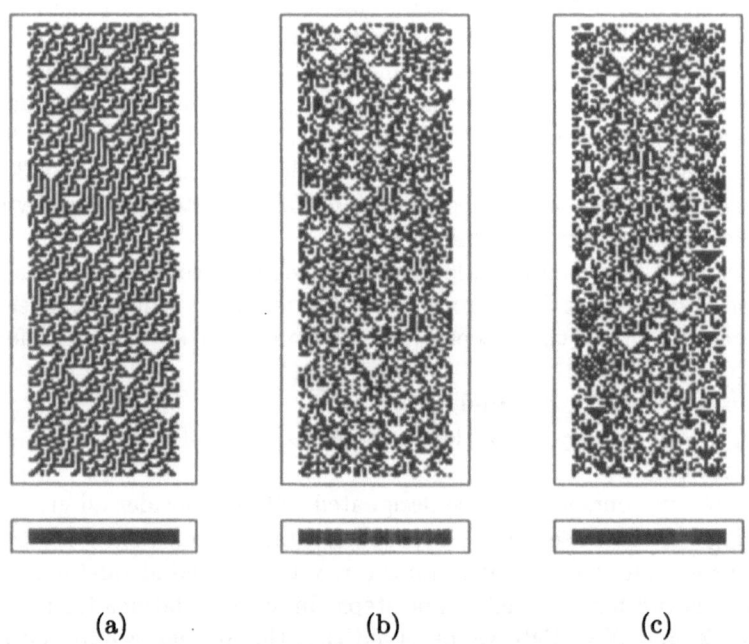

(a)                         (b)                         (c)

Figure 3: One-dimensional random number generators: Operation of three CAs. Grid size is $N = 50$, radius is $r = 1$. Top figures depict space-time diagrams, bottom figures depict rule maps. (a) Rule 30 CA. (b) Rules {90, 150} CA. (c) A co-evolved, non-uniform CA, consisting of three rules: rule 165 (22 cells), rule 90 (22 cells), rule 150 (6 cells).

# 4  Conclusions

A major impediment preventing ubiquitous computing with CAs stems from the difficulty of utilizing their complex behavior to perform useful computations. We presented the cellular programming algorithm for co-evolving computation in non-uniform CAs, demonstrating that high performance systems can be evolved for non-trivial computational tasks. Several possible avenues of research suggest themselves; one of these concerns a detailed investigation into the robustness of our systems, as described in Section 3.1. Another study which we have undertaken involves a modified model, in which the concomitant evolution of cellular rules and cellular connections takes place. We found that performance can be markedly increased for global computational tasks by such co-evolving architectures [Sipper and Ruppin, 1996a, Sipper and Ruppin, 1996b].

Evolving, non-uniform CAs hold potential for studying phenomena of interest in areas such as complex systems, artificial life and parallel computation. This work has shed light on the possibility of computing with such CAs, and demonstrated the feasibility of their programming by means of co-evolution. We believe that cellular programming holds potential for attaining 'evolving ware', evolware, which can be implemented in software, hardware, or other possible forms, such as bioware.

# References

[Chowdhury et al., 1995] D. R. Chowdhury, I. S. Gupta, and P. P. Chaudhuri. A low-cost high-capacity associative memory design using cellular automata. *IEEE Transactions on Computers*, 44(10):1260–1264, October 1995.

[Cohoon et al., 1987] J. P. Cohoon, S. U. Hedge, W. N. Martin, and D. Richards. Punctuated equilibria: A parallel genetic algorithm. In J. J. Grefenstette, editor, *Proceedings of the Second International Conference on Genetic Algorithms*, page 148. Lawrence Erlbaum Associates, 1987.

[Das et al., 1995] R. Das, J. P. Crutchfield, M. Mitchell, and J. E. Hanson. Evolving globally synchronized cellular automata. In L. J. Eshelman, editor, *Proceedings of the Sixth International Conference on Genetic Algorithms*, pages 336–343, San Francisco, CA, 1995. Morgan Kaufmann.

[Gacs, 1985] P. Gacs. Nonergodic one-dimensional media and reliable computation. *Contemporary Mathematics*, 41:125, 1985.

[Hortensius et al., 1989a] P. D. Hortensius, R. D. McLeod, and H. C. Card. Parallel random number generation for VLSI systems using cellular automata. *IEEE Transactions on Computers*, 38(10):1466–1473, October 1989.

[Hortensius et al., 1989b] P. D. Hortensius, R. D. McLeod, W. Pries, D. M. Miller, and H. C. Card. Cellular automata-based pseudorandom number generators for built-in self-test. *IEEE Transactions on Computer-Aided Design*, 8(8):842–859, August 1989.

[Knuth, 1981] D. E. Knuth. *The Art of Computer Programming: Volume 2, Seminumerical Algorithms*. Addison-Wesley, Reading, MA, second edition, 1981.

[Koza, 1992] J. R. Koza. *Genetic Programming*. The MIT Press, Cambridge, Massachusetts, 1992.

[Lamport and Lynch, 1990] L. Lamport and N. Lynch. Distributed computing: Models and methods. In J. Van Leeuwen, editor, *Handbook of Theoretical Computer Science, Volume B: Formal Models and Semantics*, pages 1159–1199. Elsevier, 1990.

[Manderick and Spiessens, 1989] B. Manderick and P. Spiessens. Fine-grained parallel genetic algorithms. In J. D. Schaffer, editor, *Proceedings of the Third International Conference on Genetic Algorithms*, page 428. Morgan Kaufmann, 1989.

[Mitchell et al., 1994] M. Mitchell, J. P. Crutchfield, and P. T. Hraber. Evolving cellular automata to perform computations: Mechanisms and impediments. *Physica D*, 75:361–391, 1994.

[Mitchell, 1996] M. Mitchell. *An Introduction to Genetic Algorithms*. MIT Press, Cambridge, MA, 1996.

[Packard, 1988] N. H. Packard. Adaptation toward the edge of chaos. In J. A. S. Kelso, A. J. Mandell, and M. F. Shlesinger, editors, *Dynamic Patterns in Complex Systems*, pages 293–301. World Scientific, Singapore, 1988.

[Park and Miller, 1988] S. K. Park and K. W. Miller. Random number generators: Good ones are hard to find. *Communications of the ACM*, 31(10):1192–1201, October 1988.

[Pries et al., 1986] W. Pries, A. Thanailakis, and H. C. Card. Group properties of cellular automata and VLSI applications. *IEEE Transactions on Computers*, C-35(12):1013–1024, December 1986.

[Sanchez and Tomassini, 1996] E. Sanchez and M. Tomassini, editors. *Towards Evolvable Hardware*, volume 1062 of *Lecture Notes in Computer Science*. Springer-Verlag, Berlin, 1996.

[Sipper and Ruppin, 1996a] M. Sipper and E. Ruppin. Co-evolving architectures for cellular machines. *Physica D*, 1996. (to appear).

[Sipper and Ruppin, 1996b] M. Sipper and E. Ruppin. Co-evolving cellular architectures by cellular programming. In *Proceedings of IEEE Third International Conference on Evolutionary Computation (ICEC'96)*, pages 306–311, 1996.

[Sipper and Tomassini, 1996a] M. Sipper and M. Tomassini. Co-evolving parallel random number generators. In H. -M. Voigt, W. Ebeling, I. Rechenberg, and H. -P. Schwefel, editors, *Parallel Problem Solving from Nature IV (PPSN IV)*, volume 1141 of *Lecture Notes in Computer Science*, Heidelberg, 1996. Springer-Verlag.

[Sipper and Tomassini, 1996b] M. Sipper and M. Tomassini. Generating parallel random number generators by cellular programming. *International Journal of Modern Physics C*, 7(2):181–190, 1996.

[Sipper, 1994] M. Sipper. Non-uniform cellular automata: Evolution in rule space and formation of complex structures. In R. A. Brooks and P. Maes, editors, *Artificial Life IV*, pages 394–399, Cambridge, Massachusetts, 1994. The MIT Press.

[Sipper, 1995a] M. Sipper. Quasi-uniform computation-universal cellular automata. In F. Morán, A. Moreno, J. J. Merelo, and P. Chacón, editors, *ECAL'95: Third European Conference on Artificial Life*, volume 929 of *Lecture Notes in Computer Science*, pages 544–554, Berlin, 1995. Springer-Verlag.

[Sipper, 1995b] M. Sipper. Studying artificial life using a simple, general cellular model. *Artificial Life Journal*, 2(1):1–35, 1995. The MIT Press, Cambridge, MA.

[Sipper, 1996a] M. Sipper. Co-evolving non-uniform cellular automata to perform computations. *Physica D*, 92:193–208, 1996.

[Sipper, 1996b] M. Sipper. Complex computation in non-uniform cellular automata, 1996. (submitted).

[Starkweather et al., 1991] T. Starkweather, D. Whitley, and K. Mathias. Optimization using distributed genetic algorithms. In H. -P. Schwefel and R. Männer, editors, *Parallel Problem Solving from Nature*, volume 496 of *Lecture Notes in Computer Science*, page 176, Berlin, 1991. Springer-Verlag.

[Tanese, 1987] R. Tanese. Parallel genetic algorithms for a hypercube. In J. J. Grefenstette, editor, *Proceedings of the Second International Conference on Genetic Algorithms*, page 177. Lawrence Erlbaum Associates, 1987.

[Thi, 1991] Thinking Machines Corporation, Cambridge, Massachusetts. *The Connection Machine: CM-200 Series Technical Summary*, June 1991.

[Tomassini, 1993] M. Tomassini. The parallel genetic cellular automata: Application to global function optimization. In R. F. Albrecht, C. R. Reeves, and N. C. Steele, editors, *Proceedings of the International Conference on Artificial Neural Networks and Genetic Algorithms*, pages 385–391. Springer-Verlag, 1993.

[Tomassini, 1995] M. Tomassini. A survey of genetic algorithms. In D. Stauffer, editor, *Annual Reviews of Computational Physics*, volume III, pages 87–118. World Scientific, 1995. Also available as: Technical Report 95/137, Department of Computer Science, Swiss Federal Institute of Technology, Lausanne, Switzerland, July, 1995.

[Vichniac et al., 1986] G. Y. Vichniac, P. Tamayo, and H. Hartman. Annealed and quenched inhomogeneous cellular automata. *Journal of Statistical Physics*, 45:875–883, 1986.

[Wolfram, 1983] S. Wolfram. Statistical mechanics of cellular automata. *Reviews of Modern Physics*, 55(3):601–644, July 1983.

[Wolfram, 1986] S. Wolfram. Random sequence generation by cellular automata. *Advances in Applied Mathematics*, 7:123–169, June 1986.

# A High-Level Language for Programming Cellular Algorithms on Parallel Machines

Giandomenico Spezzano, Domenico Talia
ISI-CNR
c/o DEIS, Università della Calabria
87036 Rende (CS), Italy

## Abstract

This paper describes CARPET, a parallel programming language based on the cellular automata model. A CARPET implementation has been used for programming cellular algorithms in the CAMEL parallel environment. CAMEL is an environment designed to support the development of high performance applications in science and engineering. It offers the computing power of a highly parallel computer, hiding the architecture issues from a user. By CARPET a user might write programs to describe the actions of thousands of simple active agents interacting locally, then the CAMEL system allows a user to observe the global complex evolution that arises from all the local interactions.

## 1 Introduction

Currently available high performance computing systems can be exploited to efficiently support applications in science and engineering. However, the lack of high-level languages, tools, and application-development environments does not allow to program parallel algorithms that are portable, efficient and expressive.

According to the Skillicorn's classification [1] the *restricted-computation structures* represent one of the most important models of parallel processing. The interest for this model is due to the possibility to restrict the form of computations so as to restrict communication volume achieving high performance. This allows to offer a user a structured model of parallel programming and improve the performance of the parallel algorithms reducing the overheads due to the communication *latency*. Further, tools can be designed to estimate the performance of various constructs of a high-level language on a specific parallel architecture.

Cellular processing languages based on cellular automata model represent a significative example of restricted-computation that it is used to model parallel computation for a large number of applications in biology, physics, geophysics, chemistry, economics, artificial life, and engineering.

A cellular automaton consists of one-dimensional or multi-dimensional lattice of *cells*, each of which is connected to a finite neighbourhood of cells which are nearby in the lattice. Each cell in the regular spatial lattice can take any of a finite number of discrete state values. Time is discrete, as well, and at each time step all the cells in the lattice are updated by means of a local rule called *transition function*, which determines the cell's next state based upon the states of its neighbors. That is, the state of a cell at a given time depends only on its own state in the previous time step and the states of its nearby neighbors at the previous time step. Different neighborhoods can be defined for the cells. The most common neighborhoods in the two-dimensional case are the von Neumann neighborhood consisting of the North, South, East, West neighbors and the Moore neighborhood composed of 8 neighbor cells. In the three dimensional case up to 26 neighbours can be taken in consideration. All cells of the automaton are updated synchronously. The global behaviour of the system is determined by the evolution of the states of all cells as a result of multiple interactions.

In our approach a cellular algorithm is composed of all the transition functions of cells that compose the lattice. Each transition function generally contains a same local rule, but it is also possible to define some cells with different transition functions (inhomogeneous cellular automata). Differently from early cellular approaches where cell state is defined as a single or a set of bits, and for extending the range of applications to be programmed by cellular algorithms we define the state of a cell as a set of typed substates that can be *shorts*, *integers*, *floats* and *doubles*. Further, we introduce a *logic neighbourhood* that inside the same radius may represent a wide range of different neighborhoods also time-dependent.

We have implemented these features in a high-level language, called CARPET (CellulAR Programming EnvironmenT), that allows to describe cellular algorithms. In particular, CARPET has been used for programming cellular algorithms in the CAMEL (Cellular Automata environMent for systEms ModeLing) [2] [3] environment. CAMEL is a software environment designed to support the parallel execution of cellular algorithms, the visualization of the results, and the monitoring of the program execution. The parallel execution of cellular algorithms is implemented by the parallel execution of the transition function of each cell in a SPMD fashion. It offers the computing power of a highly parallel computer, hiding the architecture issues from a user. By CARPET a user might write cellular programs to describe the actions of thousands of simple active agents interacting locally, then the CAMEL system allows a user to observe the global complex evolution that arises from all the local interactions.

A number of cellular programming languages such as CELLANG [4], CDL [5], CARP [6], CEPROL [7] have been defined. However none of those contains all the features of CARPET neither a parallel run-time support for them has been implemented.

Currently, CARPET is used in the CABOTO project within the PCI ESPRIT framework. The CABOTO main objective is the use of CA models for the simulation of*bioremediation* of contamined soils [8]. In this paper we briefly describe the CAMEL system and then we illustrate the full features of the CARPET language.

# 2 Overview of CAMEL

The CAMEL system is a parallel environment based on the cellular automata model for developing scientific applications. CAMEL has been implemented on a parallel computer composed of a mesh of 32 Transputers connected to a host node. The current implementation of CAMEL does not limit the number of Transputers which can compose the parallel computer, so no changes should be necessary in the software of CAMEL whether a very large number of Transputer should be used.

The CAMEL system is composed by a set of *macrocell* processes each running on a single processing element of the parallel machine and by a *controller* process running on a master processor. Each *macrocell* process implements several elementary cells and makes use of a communication system which handles the data exchange among cells.

Using this parallel architecture, CAMEL allows the parallel execution of the transition function of cells. The CAMEL system allows, by the Graphical Interface (GI), to rapidly and interactively explore and analyse very large amounts of scientific data gathered during the execution of computer simulations.

A tool called *IVT* (*Interactive Visualization Tool*), designed by MATLAB, has been added to CAMEL to improve data visualization. Utilizing data computed by simulation, *IVT* provides a variety of functions and services, including 2 and 3-dimensional graphical displays of data, hard copy of graphical displays and text, interactive colour manipulation, animation creation and display, rotation of the images, saving of data in files according to different data formats.

# 3  The CARPET language

The CARPET language is a programming tool that allows the definition of cellular algorithms. CARPET is a high-level language based on C with some additional constructs to describe the rules of the transition function of a single cell of a cellular automaton. The main features of CARPET are the possibility to describe the state of a cell as a set of typed substates each one by a user-defined type, and the simple definition of complex neighborhoods (e.g., hexagonal, Margolus, etc), that can be also time dependent, in a $n$-dimensional discrete cartesian space.

The language does not provide statements to configure the automata, to visualize the cell values or to define data channels that can connect the cells according to different topologies.

The configuration of a cellular automata is defined by the User Interface (UI) of the run-time support (i.e., the CAMEL environment). The UI allows, by menu pops, to define the size of the cellular automata, the number of the processors onto which the automata must be executed, and to choose the colours to be assigned to the cell substates to support the graphical visualization of their values.

The exclusion from the language of constructs for configuration and visualization of the data allows to execute the same CARPET program with different configurations. Further, it is possible to change from time to time the size of the automaton and/or the number of the nodes onto which the automaton should be executed. Finally, this approach allows to select the more suitable range of the colours for the visualization of data. The execution of a program with different configurations allows to evaluate a model with different partitions of the space, obtained changing the size of the cell. This allows also to measure the scalability and the efficiency of the system.

Using CARPET a wide variety of cellular programs can be described in a simple but very expressive way. The language utilizes control structures, types, operators and expressions of the C language. However it is enhanced by a declaration part that allows to specify the dimensions of the automaton, the radius of the neighbourhood, the type of the neighbourhood, and to describe the state of a cell as a set of typed substates that can be: *shorts*, *integers*, *floats* and *doubles*. Furthermore, a set of global parameters describe the global characteristics of the system (e.g., the permeability of a soil).

Special functions allow to modify, at each iteration, the values of the substates of a cell and to define a set of cells (e.g., those of the border) with a different transition function. This last characteristic is very interesting because it simplifies the modeling phase of a system that can be represented by a network of cellular automata each describing one of the components in which has been divided the model to capture the different aspects of a phenomenon.

The structure of a CARPET program is similar to that of a C program. A CARPET program is composed by a *declaration part* that appear only once in the program and must precede any statement (except those of C pre-processor) and by a *body program*. The *body program* has the usual C statements, without I/O instructions, and a set of special functions to modify the state of a cell and its neighborhood. CARPET allows to use C functions or procedures to improve the structure of the programs.

## 3.1 Declaration part

The declaration part describes the dimensions of an automaton, the radius of the neighbourhood, the state of cells, the cells belonging to the neighborhood and the global parameters. These declaration are contained inside of the **cadef** section.

```
cadef
{
  dimension n;
  radius m;
  state { type_specifier substate_name, substate_name;

          type_specifier substate_name .... }

  neighbor id[n] {[xvalue, yvalue, zvalue] id, ...
                    [xvalue, yvalue, zvalue] id };
  parameter {id value, id value, .....};

}
```

### 3.1.1 Dimension

This definition allows to specify by a numeric literal the number of dimensions of an automaton, in a discrete cartesian space. For example:

**dimension** 3

defines a three-dimensional automaton.

The example above shows the maximum number of dimensions allowed in our implementation. Each dimension is wrap around, e.g., a two-dimensional lattice forms a torus. Border functions can be used to disable the wrap-around.

### 3.1.2 Radius

Radius defines a numeric value that specifies the radius of the neighborhood of a cell. This value is strictly connected with the **dimension** definition. For example, in a 2-dimension automaton defining the radius equal to 1 the number of the neighbours can be up to 8. In the three dimensional case with radius equal to 1 the number of the neighbours can be up to 26. The next example defines a radius equal to 2:

**radius** 2

Our implementation supports a radius equal to 1 for the three dimensional case, up to 2 for the two dimensional case, and up to 50 for the one dimensional case.

### 3.1.3 State

In CARPET the state of a cell is constituted by a set of typed substates, unlike classical cellular automata where the state is represented by a few bits. The types of substates are: *shorts* (16 bits integers), *integers*, *floats* (reals), *doubles* (64 bits reals).

By typification of substates, CARPET allows to extend the range of the applications that can be coded by cellular algorithms simplifying the writing of the programs and improving their readability.

Most systems and languages such as CELLANG, define the cell substates only as *integers*. In this case, for example, if a user must store a real value in a substate then he must write some procedures for the data retyping. The writing of these procedures makes the program longer and difficult to read or change. The CARPET language frees the user of this tedious task and offers him a high level abstraction to define the cell state. The set of substates is defined through the **state** declaration. All substates are contained in round brackets and separated by commas. A type specifier must be included for each substate. In the following example the state is constituted of three substates.

**state**(**short** temp, quote; **float** volume);

The first two substates *temp* and *quote* are shorts, the third *volume* is a float.
The predefined variable **cell** refers the current cell in the n-dimensional space under consideration. The different substates can be referred appending to the reserved word

**cell** the name of the substate by the underscore symbol ( _ ). For example, **cell_volume** refers the *volume* substate of the previous example.

### 3.1.4 Neighbor

As mentioned before, through the radius it is possible to define the maximum number of cells which can compose the neighborhood of a cell and that can be accessed to read their state.

To read a substate of a neighbor cell is needed to specify the indexing of the cell relative to the current cell. Relative indexing is explicited by a number of indexes equal to the dimension of the automata enclosed in square brackets and separated by commas. The figure 1 shows, for a two dimensional automata with radius equal to 1, the indexing of a cell relative to the current cell indexed with [0,0]. For example, the cell **N** situated at north is indexed with [0,-1], having used in the implementation the reference system indicated in figure 1.

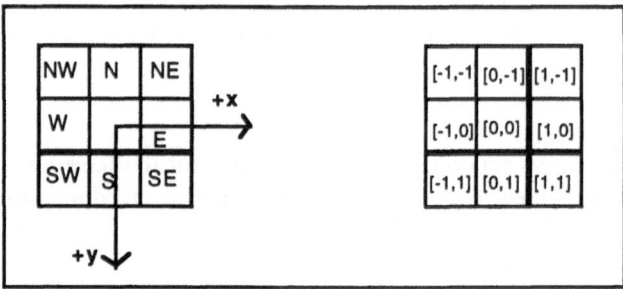

**Figure 1.** Relative indexing of a cell with respect to the cell [0,0].

CARPET generalizes the concept of neighbourhood and allows the user to define by the **neighbor** declaration a *logic neighbourhood* that inside the same radius may represent a wide range of different neighborhoods. Neighborhoods can be asymmetrical or have any other special topological properties (e.g., hexagonal).

The neighborhood is identified by the name of a vector with dimension equal to the number of elements composing the logic neighbourhood. The elements of the logic neighbourhood are put in round brackets and are separated by commas. Each element is described by relative indexing. Furthermore, to each of these elements it can be associated a name that can be used as an alias in referring to the neighbour cell. The von Neumann neighbourhood can be defined as follows:

```
neighbor Neumann[4] ([0,-1] North, [-1,0] West,
                     [0,1] South, [1,0] East);
```

The example of figure 2 shows how through the **Neumann** vector it is easy to access to the **green** substate of the cell situated at **West**. This access way simplifies to access the substates using the **for** statement. An alternative to access **green** substate of the cell located at West is **West_green**.

By the predefined variable **step** CARPET allows to know the number of iterations that have been executed. **Step** is updated automatically by the system. Initially the value of **step** is 0 and it is incremented by 1 each time that all the cells of the automaton have been updated. This feature allows also to implement neighborhoods which are time dependent.

```
cadef
{
    dimension 2;
    radius 1;
    state (short green, red);
    neighbor Neumann[4] ([0,-1] North, [-1,0]West,
                         [0,1]South,[1,0] East);
}
    short value;
    value = Neumann[1]_green;
```

**Figure 2.** Referencing the **green** substate of a neighbour cell.

**Step** allows also to change dynamically the values of the substates dependent upon the iterations. A system characterized by several temporal phases can be described using this feature in a transition function. For example, the first iteration can be used to set up some parameters, the second iteration can initialize some substates and the next iterations can calculate, for a defined number of steps, the transition function describing the phenomenon.

*3.1.5 Global parameters*

In modeling a complex system it is often necessary to describe some global features of the system. CARPET allows to define global parameters and to initialize them to specific values. The value of a parameter is the same in each cell of the automaton. For this reason, the value of each parameter cannot be changed in the program but it can only be modified, during the simulation, by the UI. The global parameters can be declared by the **parameter** construct. To each global parameter is associated a name used to refer the parameter during the computation. The type of parameters must be **float**, by default its value is zero.The following example shows the use of two global parameters **adherence** and **permeability** initialized to 0.5 and 10.0.

> **parameter** (adherence 0.5, permeability 10.0);

```
cadef
{
 ......
 parameter (adherence 0.5);
}
if (adherence < 0.7)
    sum = sum + quote;
else
  sum  = temp - cdiff;
 ......
```

**Figure 3.** Use of the **parameter** definition.

The example in figure 3 shows as a parameter can be used with the usual conditional statement **if**.

## 3.2  Statements

To guarantee the semantics of cell state updating in cellular automata the value of one of the substates of a cell can be modified only by the **update** function. After an **update** statement the value of the substate, in the current iteration, is unchanged.

The new value does not take effect until the beginning of the next iteration. For this reason, the updating of the value of a substate cannot be performed by a direct assignment. The output of the program will be wrong if a value is assigned to a substate without the **update** statement. For example, the function:

```
update (cell_temp,45);
```

assigns to the temp substate the value 45. This value will be really available only in the next iteration. Input and output of a CARPET program can be performed by files o by the edit function of the UI. A file can contain the values of one substate of all cells, these values can be loaded at the step 0 to initialize the automaton. These values can be the result of a previous simulation or they can be generated by a C program. In fact, the output of a CARPET program can be used as input to initialize another automaton because the format of the input and output is identical.

In regular intervals the output of a CARPET program can automatically be saved in a file to calculate global statistical functions (i.e. histogram, etc) or to be post-processed by a visualization tool. A user can use additional substates to store values that indicate statistical proprieties of a variable (i.e. mean ) or to hold a history of a substate. For instance, the average of the **temp** substate can be calculated and stored as shown in figure 4. Notice that the predefined **step** variable indicates the number of iterations that have been executed.

```
cadef
{
   dimension 2
   radius 1
   state ( short x, y; float temp, mean);
   ....
}
   ....
   avgtemp = (cell_mean + cell_temp ) / step;
   update(cell_mean, avgtemp);
```

**Figure 4.**  Evaluation of the average of the temp substate.

### 3.2.1 Special functions

The rules defined in CARPET are *deterministic,* i.e., the new state of a cell is uniquely determined by the current state of its neighbors: from the same initial conditions one invariably obtains the same evolution. However, CARPET offers the possibility to define non deterministic rules by the use of a *random* function.

CARPET allows to define cells with different transition functions by means of the **GetX, GetY, GetZ** functions that return the value of the coordinates X,Y, and Z of the cell in the automaton. Using those functions it is possible to specify a different transition function for a single cell. Varying only a coordinate it will be possible to associate the same transition function to all cells belonging to the same row or column. The example in figure 5 shows how to assign a different transition function for the cell having coordinates (5,8).

```
cadef
{
   dimension 2;
   radius 1;
   ....
}
 Xpos = GetX;
 Ypos = GetY;
 if (Xpos == 5 && Ypos == 8)
    func_trans_1();
  else
    func_trans_2();
```

**Figure 5.** Definition of a different transition function for the cell (5,8).

## 4 An example

The example in figure 6 shows how the CARPET language can be used to implement a simple simulation of the propagation of a forest fire.

```
#define ground 0
#define fire 1
#define tree 2
cadef
{
 dimension 2;     /*bidimensional lattice */
 radius 1;
 state (short land);
 neighbor cross[4] ([0,-1]North,[-1,0]West, [0,1]South,[1,0]East);
}
{
  if (cell_land == fire) ||
     (cell_land == tree &&
     (North_land == fire || South_land == fire ||
      East_land == fire || West_land == fire)
     update(cell_land, cell_land - 1)/* the new state of cell*/
}
```

**Figure 6.** The forest fire simulation written in CARPET.

In this example the cells can have the values included between '0' and '2'. The ground is represented by '0' value, the fire is represented by '1' value and the tree is represented by '2' value. Each cell represents a portion of the land. Fire spreads from a cell which is on fire to a von Neumann neighbor that is treed, but not on fire.

# 5 Conclusions

Our experience during the design, implementation and use of the CARPET language showed us that high-level languages are very useful for the development of cellular algorithms for solving complex problems in science and engineering. The CARPET approach is quite different from that followed in the implementation of early cellular processing systems where low-languages are used to implement cellular algorithms.

These languages make difficult to implement, read and port to other systems the cellular algorithms. In our opinion high-level languages as CARPET will allow to enlarge the use of cellular automata in solving complex problems preserving high performance and expressiveness.

## Acknowledgements

This research is partially supported by CEC ESPRIT P.C.I. contract n° 94529419370.

## References

1. Skillicorn D B. Foundations of parallel programming, Cambridge Series in Parallel Computation, Cambridge University Press, 1994

2. Cannataro M, Di Gregorio S, Rongo R, Spataro W, Spezzano G, Talia D. A parallel cellular automata environment on multicomputers for computational science. Parallel Computing 1995; 21:803-824

3. Di Gregorio S, Rongo R, Spataro W, Spezzano G, Talia D. A parallel cellular tool for interactive modeling and simulation. IEEE Computational Science & Engineering 1996; 3,3:33-43

4. Eckart J D. Cellang 2.0: reference manual. ACM Sigplan Notices 1992; 27, 8:107-112

5. Hochberger C, Hoffmann R. CDL- a language for cellular processing. Proc. 2nd Intern. Conference on Massively Parallel Computing Systems, IEEE Computer Society 1996

6. Junger G. Cellular automaton tool user manual. GMD, Sankt Augustin 1994

7. Seutter F. CEPROL a cellular programming language. Parallel Computing 1985; 2:327-333

8. Di Gregorio S, Rongo R, Serra R, Spataro W, Villani M. Simulation of water flow through a porus soil by a cellular automaton model. these proceedings

# AUTHOR INDEX